Lean Healthcare Deployment and Sustainability

Mark L. Dean, Ph.D.

New York   Chicago   San Francisco   Lisbon   London
Madrid   Mexico City   Milan   New Delhi   San Juan
Seoul   Singapore   Sydney   Toronto

Cataloging-in-Publication Data is on file with the Library of Congress.

McGraw-Hill Education books are available at special quantity discounts to use as premiums and sales promotions, or for use in corporate training programs. To contact a representative please e-mail us at bulksales@mcgraw-hill.com.

Lean Healthcare Deployment and Sustainability

Copyright ©2013 by McGraw-Hill Education. All rights reserved. Printed in the United States of America. Except as permitted under the United States Copyright Act of 1976, no part of this publication may be reproduced or distributed in any form or by any means, or stored in a data base or retrieval system, without the prior written permission of the publisher.

1 2 3 4 5 6 7 8 9 0 DOC/DOC 1 9 8 7 6 5 4 3

ISBN 978-0-07-181770-7

MHID   0-07-181770-0

The pages within this book were printed on acid-free paper.

Sponsoring Editor	**Proofreader**
Judy Bass	Claire Splan
Acquisition Coordinator	**Indexer**
Amy Stonebraker	Jack Lewis
Editorial Supervisor	**Production Supervisor**
David E. Fogarty	Pamela A. Pelton
Project Manager	**Composition**
Patricia Wallenburg	TypeWriting
Copy Editor	**Art Director, Cover**
Patti Scott	Jeff Weeks

Information contained in this work has been obtained by McGraw-Hill Education LLC from sources believed to be reliable. However, neither McGraw-Hill Education nor its authors guarantee the accuracy or completeness of any information published herein, and neither McGraw-Hill Education nor its authors shall be responsible for any errors, omissions, or damages arising out of use of this information. This work is published with the understanding that McGraw-Hill Education and its authors are supplying information but are not attempting to render engineering or other professional services. If such services are required, the assistance of an appropriate professional should be sought.

To my darling Bev

About the Author

Mark L. Dean, Ph.D., is an innovative thought leader and seasoned executive with over 30 years' experience in healthcare, government, academia, and industry. He holds a doctorate in clinical psychology and master's degrees in electrical engineering and management. Dr. Dean has successfully guided healthcare, engineering, manufacturing, and service organizations in their pursuits of organization excellence by facilitating the development and implementation of leadership systems that integrate strategic, operational, and quality improvement initiatives. He is the author of *Healing Healthcare: A Leadership Journey* (Goal/QPC, 2012) and the *Lean Memory Jogger for Healthcare* (Goal/QPC, 2012).

CONTENTS

PART I

Lean Leadership

PART II

Healing Enterprise Transformation

PART III

Healing Pathway Transformation

PART IV

Process Transformation

PART V

Continual Strategic and Operational Improvement

INTRODUCTION

This book provides a comprehensive guide for implementing Lean methodologies in a hospital, physician practice, long-term care facility, or any other healthcare setting. It is intended to be used by those charged with the responsibility for leading Lean initiatives and organizational transformations. It is also useful as an instruction manual for the teaching of such initiatives.

The clear success of Lean in other industries has stimulated its adoption in healthcare. A growing number of progressive healthcare organizations are reaping its benefits. These include the Mayo Clinic, ThedaCare, Jewish Hospital, Virginia Mason, and Boston Medical Center. The success of these pioneers demonstrates the improvement results available to those who apply Lean lessons and serves as a model for all healthcare organizations.

Healing Pathways

This book introduces the concept of *healing pathways*, which are defined as value streams through which patients flow. A *value stream* is a set of one or more processes that result in the delivery of a product or service. *Value stream analysis* is the study and improvement of value streams and is the foundation upon which Lean initiatives are based. Its objective is to improve the flow of products and services through the value stream by eliminating waste and increasing value. For the most part, a value stream analysis largely ignores the perspective of the product or service that flows through it. In production and most service processes, that is an appropriate approach. However, in a healthcare setting, such an approach may not place enough emphasis on the patient.

Patients, unlike widgets, perceive their environment and respond to it. To facilitate the health and well-being of patients, not to mention their satisfaction, there must be a greater focus on the fact that they are living, perceiving beings who interact with their environment.

Therefore, rather than continue to use the term *value stream*, this book uses the term *healing pathway* when a patient is involved in the process. In practice, healthcare organizations may continue to use the term *value stream* for support processes and the delivery of other products and services. However, to emphasize the primacy of the patient, and a focus on the patient's perception and experience in every improvement activity, this book relies heavily on the concept and terminology of healing pathways, and may use the terms *value stream* and *healing pathway* interchangeably.

Healthcare is a healing enterprise. Healing pathways are the core of the healthcare industry. Healing pathways, and not value streams, take primacy.

Introduction to the Lean Healthcare Transformation Process

Enterprise transformation is a monumental undertaking. To be successful, the CEO and the senior leadership team need to be fully engaged and committed. They need to be visible leaders, cheerleaders, and enablers of the transformation initiative. Lean, in and of itself, does not create organizational transformation. It can be successful only when employed as a tool by the leaders at all levels of the enterprise in a cascading cycle of improvement and oversight activities that start at the top. Enterprise transformation requires three interdependent levels of activity, as depicted in Figure I.1:

▲ *Enterprise transformation* focuses on the healthcare services and enterprisewide systems, while directing specific value stream transformations. It must be led by the chief executive officer and the senior leadership team.

▲ *Healing pathway transformation* focuses on high-level macro processes that impact either a healing pathway (clinical process) or value stream (nonclinical process). These must be led by the senior executive over the respective area.

▲ *Process transformation* focuses on the specific processes that combine to form the healing pathways. Process transformations must be led by the senior leader over the respective area.

At each level of the transformation, Lean tools and methodologies are embedded in an iterative PDCA (Plan-Do-Check-Act) cycle, where each

Figure I.1 The Lean healthcare transformation process.

level is responsible for setting the priorities and ensuring the success of the efforts subordinate to it. This book provides guidance on how to use the Lean tool kit. More than that, it provides an organizational infrastructure and approach for achieving transformation of your healing enterprise. It is a comprehensive guide for enterprise transformation in the pursuit of an unachievable yet worthy goal: perfect care.

Healing University Healthcare System Examples

This book uses numerous examples and case studies to support explanations of various organizational transformation activities. The examples, case studies, and results are real, and come from the author's work in various healthcare organizations. In this book they are presented using Healing University Healthcare System, a fictional but representative system. Healing Univserity includes a main hospital, two ambulatory facilities, a long-term care facility, and a small but increasing number of physician practices. A companion text entitled *Healing Healthcare: A Leadership Journey* presents a fictional narrative that follows the journey of Healing University as it transforms itself through the application of the Lean principles and practices provided in this book. *Healing Healthcare* is targeted to executives and senior leaders and is intended to provide a quick but compelling case for Lean in healthcare. This present book provides the "how."

How to Use This Book

This book will define and explain the process and the tools you need to transform your healthcare system into a Lean enterprise. The specific information provided includes the following:

▲ A transformation road map
▲ Implementation tools
▲ Specific steps to take

This information will help you and your team work together to systematically achieve your Lean enterprise goals.

ACKNOWLEDGMENTS

My sincere thanks to the people and organizations who reviewed the draft versions of this manuscript and offered suggestions and encouragement. Their participation in the development process assured that the methods and tools described are relevant and appropriate for all organizational change leaders and instructors to use in their quest to achieve a truly Lean healthcare enterprise.

Jennifer Dean
Advanced Endodontics

Chuck DeLadurantey
Community Health Systems

Michael McIntosh
Catholic Health Initiatives, Inc.

Floyd McKeag
Corio Consulting, LLC

Mike Mudd
Jewish Hospital & St. Mary's HealthCare

Val Slayton, MD, MPP, MBA, CPE
Jewish Hospital & St. Mary's HealthCare

Robert Strickland
Catholic Health Initiatives, Inc.

Laura Swessel
Humana

Glenn Whitfiled
Signature HealthCARE, LLC

Jeff Zachary
Humana

PART I

Lean Leadership

CHAPTER 1

The Business Case for Lean in Healthcare

What Is Lean?

Lean is a process improvement approach that focuses on increasing value by eliminating waste and increasing the throughput of customer-driven value streams. Traditionally, this means reducing the time between the presentation of a need, in this case a patient, and product or service delivery, in this case patient care. It is useful to consider improved flow as a desired outcome. Flow is especially helpful in healthcare settings where we can consider the flow of patients through the admission process, emergency department, radiology, inpatient care, and every other patient process. A Lean approach in healthcare achieves process flow and improved care by focusing on eliminating waste in the process and by eliminating impediments to healing.

Toyota Motor Company initially developed this disciplined, process-focused production system, implementing the Toyota Production System, now known as *Lean production* or *Lean enterprise*. Lean (the Lean enterprise concept) was popularized in U.S. manufacturing through a Massachusetts Institute of Technology study of the Japanese automotive industry, described in *The Machine That Changed the World* (J. Womack, D. Jones, and D. Roos, Free Press, NY, 1990). That book describes how Japanese business methods use less human effort, capital investment, floor space, materials, and time in all aspects of operations. The competition between U.S. and Japanese automakers has led to the successful adoption of these principles by all U.S. auto manufacturers. Lean is based on the five core principles of value, value stream mapping, pull production, flow, and seeking perfection:

▲ *Value* is defined from the perspective of the customers—it is what benefits them.

▲ *Value stream mapping* identifies both waste and value, documenting the steps necessary to transform inputs into outputs.

▲ *Pull production* produces only what is needed for the customer, when it's needed.

▲ Value *flows* after waste is eliminated and pull production is in place.

▲ *Seeking perfection* in healthcare means *perfect care!*

The primary benefits of Lean include increased productivity and throughput, reduced waste, reduced number of defects, increased quality, reduced lead times, less inventory, and consistent, on-time delivery. The manufacturing sector has adopted Lean processes. However, Lean manufacturing language cannot be seamlessly inserted into healthcare organizations without translation and guidance for its use. More importantly, in healthcare the concept of Lean must be expanded beyond the elimination of waste and increased throughput to acknowledge that what flows through many of our processes are people, not widgets. This means we must also consider the impact of our processes on our patients. Patients are the driving force behind this book and its concept of healing pathways as value streams through which patients flow.

Why Lean in Healthcare?

As healthcare administrators, as physicians, as nurses, as staff, we strive to achieve excellent patient care, delivered in an efficient and timely flow. Perhaps even more so, we want the same excellent care for ourselves and our family members. We want a healthcare system that can provide excellent care for those we love, when they are at their weakest and their most vulnerable, at a cost we can all afford.

So what keeps us from achieving perfect care? Well, let's acknowledge that healthcare is a very complex system, for reasons known to all of us:

▲ We work with ill and fragile people at some of the most challenging times of their lives.

▲ We face a myriad of laws, rules, and regulations.

▲ Physicians and hospitals are not always in alignment.

▲ Staff is frequently overworked.

▲   Reimbursements continue to decline.
▲   Costs continue to increase.

According to the Centers for Medicare and Medicaid Services (CMS), U.S. healthcare spending increased 3.9 percent in 2011. Total health expenditures reached $2.7 trillion, which translates to $8680 per person, or 17.9 percent of the nation's gross domestic product (GDP), up from 16.6 percent in 2008. Over the 2015–2021 period, CMS estimates an average annual health spending growth of 6.2 percent, anticipated to outpace average annual growth in the overall economy. By 2021, national health spending is expected to reach $4.8 trillion and comprise 19.6 percent of the GDP.

The economic and political environments we face today suggest that healthcare delivery will continue to become more, not less, challenging. The Patient Protection and Affordable Care Act (PPACA) of 2010 and the American Recovery and Reinvestment Act (ARRA) of 2009, with its Health information Technology, or HiTech, component, are just two recent examples of sweeping industry reforms. No doubt these types of reforms will continue.

Continual change in an environment of increasing complexity presents both challenges and opportunities. Organizations that can respond nimbly to improve time to market for new service delivery models have an opportunity to thrive. Those that cannot, will not survive.

This book provides a road map and the tools needed to meet the challenges of today—and tomorrow. Lean can neither predict nor control the political or economic factors outside the healthcare enterprise. What Lean *can* do is to help an organization be prepared for any eventuality. At the end of the day, being nimble, providing higher-quality care, and doing so at a lower cost are key to not only surviving, but also thriving.

Continual improvement methodologies have been used to improve quality, productivity, and competitive position in a variety of industries, including the automotive, aerospace and defense, biotechnology, and broad manufacturing industries. Successful continual improvement approaches include total quality management (TQM), ISO 9000, Six Sigma, the Malcolm Baldrige National Quality Award, and most recently Lean and Lean Sigma. These have gained wider acceptance in healthcare as pioneers have begun to reap its benefits.

There is a growing literature with case studies and examples that richly demonstrate how the implementation of Lean principles can yield dramatic improvements in healthcare:

▲ Methodist Health System in Omaha reduced uncompensated patient care from 3.4 percent to 2.5 percent (The Academy of Healthcare Revenue. *Optimizing Financial Clearance Through Lean Production Methods in Patient Access*, 2009).

▲ The University of Pittsburgh Medical Center Shadyside dramatically reduced patient falls—from an average of one fall every 12 hours to 95 days with no falls (Steven J. Spear. Fixing Healthcare From the Inside, Today, *Harvard Business Review*, Harvard Business School Publishing Corporation, Boston, 2005).

▲ The Community Medical Center in Missoula, Montana, reduced the time in the recovery room from 90 to 62 minutes, improving patients' access to care by 20 percent (Patricia Panchak. Lean Health Care? It Works! IndustryWeek.com, *Industry Week*, December 21, 2004).

▲ St. Luke's Hospital in Houston, Texas, improved patient flow, including a 5.1 percent improvement in overall capacity, 4-hour reduction in patient length of stay, and 76 percent improvement in bed turnaround time (D. Pate and M. Puffe. Improving Patient Flow, *The Physician Executive*, The American College of Physician Executives, May–June 2007).

▲ Virginia Mason reduced inventory costs by more than $1 million and reduced labor expenses in overtime and temporary labor by $500,000 in one year (Gary S. Kaplan and Sarah Patterson. Seeking Perfection in Healthcare, *Healthcare Executive*, The American College of Healthcare Executives, May/June 2008).

▲ Jewish Hospital in Louisville, Kentucky, improved the throughput of the heart catheterization lab by 28 percent and improved patient satisfaction scores by 50 percent (Ed Green. Cutting the Fat: Area Hospitals Use Lean-Management Approach to Save Money and Time, *Business First of Louisville*, April 2007).

▲ ThedaCare reduced accounts receivable from 56 to 44 days equating to about $12 million in cash flow and achieved $3.3 million in savings in one year (Diane Miller, editor. *Going Lean in Health Care*, Institute for Healthcare Improvement, Cambridge, 2005).

Dr. Donald Berwick, founder and former chief executive of the Institute for Healthcare Improvement, suggests that it's time for the healthcare industry to draw on quality improvement techniques common in other industries. This book provides a useful tool to help healthcare professionals accomplish that objective.

CHAPTER 2

The Lean Transformation Process

What Is It?

The Lean transformation process is a structured organizational approach for prioritizing and aligning strategic and operational improvements to achieve transformative benefits in patient care, operational efficiency, and physician and team member satisfaction. It is the road map for achieving "perfect care" as the aspirational goal of every healing enterprise. Deployed correctly, it will result in excellent patient care provided by competent and caring staff, nurses, and physicians, with economically sustainable efficiency.

Some Terms to Know

▲ An *enterprise map* is a high-level diagram of the services provided by the enterprise.

▲ A *transformation summit* is an offsite planning session during which the CEO and senior leadership team (SLT) plan the enterprise transformation.

▲ A *value stream* is a series of one or more processes that result in the delivery of a service or product.

▲ A *healing pathway* is a value stream through which patients flow. An example is the treatment of a patient in the emergency department.

▲ A *value stream analysis* (VSA) is a structured and facilitated event during which a cross-functional team progresses methodically through a series of interactive and evaluative steps aimed at eliminating waste and improving the flow of products or services through the value stream.

Figure 2.1 A healing pathway is a type of value stream.

▲ A *healing pathway analysis* (HPA) is a structured and facilitated event during which a cross-functional team progresses methodically through a series of interactive and evaluative steps aimed at eliminating waste, eliminating impediments to healing and improving the flow of patients through the healing pathway.

▲ A *process* is a set of steps that transform one or more inputs into one or more outputs. One simple example of a process is a nurse drawing blood from a patient.

Note that the term *healing pathway* was chosen very intentionally to emphasize a focus on the patient's perception and experience in every improvement activity. To emphasize the primacy of the patient, this book uses the term *healing pathway* interchangeably with the term *value stream*. However, for the sake of clarity, it should be understood that a healing pathway *is* a type of value stream. Specifically, it is a value stream that provides patient care, as shown in Figure 2.1.

How Do I Do It?

Lean transformation requires interdependent efforts at three cascading levels in the organization: the enterprise level, the healing pathway (value stream) level, and the process level (Figure 2.2).

▲ *Enterprise transformation* focuses on the enterprise-wide healthcare services and systems, while directing specific healing pathway and value stream transformations. The enterprise transformation must be led by the chief executive officer and the senior leadership team. It is their responsibility to direct and prioritize transformation efforts in alignment with the enterprise's strategic and operational priorities. The enterprise transformation is guided by a transformation plan that will be developed by the senior leadership team during the transformation summit.

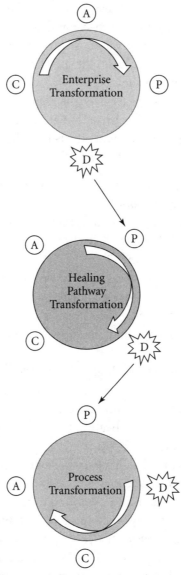

Figure 2.2   The transformation process.

▲ *Healing pathway/value stream transformation* focuses on high-level processes that constitute the healing pathways (clinical processes) and business pathways (value streams). These transformations must be led by the senior executive over the respective area. Transformation will be guided by a healing pathway analysis and rapid improvement plan that will be developed by the executive and other key leaders during a healing pathway analysis or value stream analysis event.

▲ *Process transformation* focuses on the specific processes that combine to form the healing and business pathways. Process transformations must be led by the senior manager over the respective area. Process transformations will be guided by action items developed by teams or individuals during improvement efforts including rapid improvement events, Six Sigma projects, 5S's, and others.

Lean enterprise transformation is achieved by establishing three interdependent and tightly coordinated levels of activity. Those familiar with the tenets of total quality management (TQM) will appreciate the usefulness of Deming's Plan-Do-Check-Act (PDCA) cycle as a worthy model to guide transformation.

The Plan stage of each step is focused on the activities of that level: planning the Lean enterprise transformation at the enterprise level, planning the healing pathway or value stream transformation at the healing pathway or value stream level, and planning the process transformation at the process level.

Note that the Do phase of a higher-level cycle is implemented by the complete PDCA of the supporting cycle. For example, the Do of the enterprise transformation is accomplished by a series of healing pathway PDCAs. An example of such a healing pathway is the flow of a patient through the emergency department.

Similarly, the Do of the healing pathway/value stream transformations is implemented by the complete PDCA cycle of the process transformation. Take, for example, an organization working to transform the emergency department healing pathway. Examples of potential processes to be improved include triage, physician assessment, and lab work.

The Check and Act steps of each cycle involve review of the activities under it. For example, the process transformation Check involves ensuring that all the action items identified to improve that process are being completed. The enterprise transformation Check, on the other hand, involves ensuring that each of the healing pathway/value stream transformations is proceeding as planned.

The following case study provides an example of how the three levels cascaded to achieve improvement in the lab at the Healing University Healthcare System.

CASE STUDY

Cascading Interdependence of Enterprise, Healing Pathway, and Process Level Transformations

Healing University Healthcare System

The following example, for one of Healing University's Lean improvement efforts, illustrates the cascading interdependence of enterprise, healing pathway, and process-level transformations:

▲ **At the Enterprise Level:** The CEO and senior leadership team participated in a transformation summit to prioritize and plan their enterprise transformation efforts. They prioritized a number of healing pathways for immediate transformation, including the emergency department (ED). The vice president (VP) of the Emergency Department was assigned the responsibility for leading the effort.

▲ **At the Healing Pathway Level:** The VP over the EDs participated in a healing pathway analysis to study the current state, envision a future state, and develop a plan for achieving the future state of the emergency department. The team identified a number of process improvement opportunities. One of those was the process of getting patients' lab results more quickly. The lab director was assigned responsibility for leading an effort to improve that process.

▲ **At the Process Level:** The lab director participated in a rapid improvement event to study the current state of the process by which orders for patient labs were received, processed, and fulfilled. The team identified a number of opportunities for improvement, which they documented in an action plan. These items were completed over the next couple of weeks.

Lean Enterprise Transformation

Using the PDCA cycle, the transformation at the enterprise level is comprised of the following steps, depicted in Figure 2.3:

▲ *Plan* the Lean enterprise transformation. This is primarily accomplished through the transformation summit.

   ▼ Assemble and train a leadership team.

   ▼ Identify and prioritize enterprise-level improvement opportunities that align with strategic and operational priorities.

   ▼ Assess the organizational ability to solve difficult, complex, and chronic challenges of the targeted services and processes.

   ▼ Assess resource readiness to include required capabilities and availability to conduct Lean initiatives.

▲ *Do* the enterprise transformation. The enterprise transformation is accomplished by transformation of prioritized healing and business pathways.

   ▼ Gather and direct the resources necessary to specific transformational efforts.

   ▼ Assign accountability for outcomes and responsibility for executing transformational tasks.

   ▼ Plan each of these including assigning an executive sponsor and developing a charter.

   ▼ Do, using healing pathway/value stream analyses and other appropriate improvement tools.

   ▼ Check to ensure the improved process is hardwired.

   ▼ Act to eliminate barriers to process improvements.

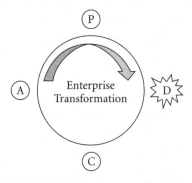

Figure 2.3 Enterprise transformation.

▲ *Check* the enterprise transformation.
   ▼ Provide executive-level oversight to ensure the enterprise transformation is achieving sustainable results.
▲ *Act* to eliminate barriers to Lean enterprise transformation.

Healing Pathway Transformation

There must be clear interdependence between enterprise and healing pathway/value stream transformational efforts. Note that the Do of the enterprise level is implemented entirely by the complete PDCA cycle at the healing pathway/value stream level. This relationship is shown in Figure 2.4. Transformation of the healing (or business) pathway is the mechanism by which the enterprise is transformed.

▲ *Plan* the healing pathway/value stream transformation. This is primarily accomplished through the healing pathway/value stream analysis.
   ▼ Formalize the team charter with the executive sponsor.
   ▼ Schedule and prepare for the healing pathway analysis (HPA) or value stream analysis (VSA) event.
   ▼ Understand the current state.
   ▼ Design the future state.
   ▼ Develop a rapid improvement plan.
▲ *Do* the transformation. Healing pathways and value streams are transformed by improvement of the processes of which they are comprised. These are prioritized in the rapid improvement plan.

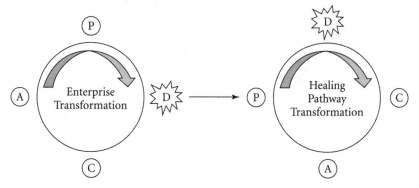

Figure 2.4   The healing pathway transformation is the Do of the enterprise transformation.

▼ Conduct rapid improvement events.
▼ Conduct Six Sigma projects.
▼ Complete the Just Do It items.
▼ Conduct 5S and other Lean improvement efforts.
▲ *Check* the healing pathway/value stream transformation.
  ▼ Provide management-level oversight to ensure the healing pathway transformation is achieving sustainable results.
▲ *Act* to eliminate barriers to the healing pathway/value stream transformation.
  ▼ Standardize new procedures.

Process Transformation

Finally, there must be clear interdependence between healing pathway/value stream and process transformation efforts. This interdependence is provided by the rapid improvement plan, the healing pathway's executive sponsor, and their review of process transformation efforts. Note that just as the Do of the enterprise transformation is achieved through the complete PDCA cycle of the healing pathway/value stream transformation, so the Do of the healing pathway/value stream transformation is achieved through execution of the complete PDCA cycle of the process transformation. This relationship is shown in Figure 2.5.

Process transformation efforts are achieved through a variety of individual and team-based events. These include Just Do Its (JDIs), rapid

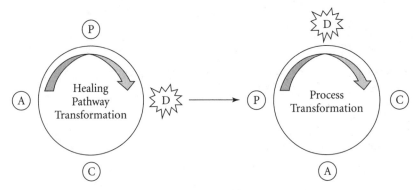

Figure 2.5 Process transformation is the Do of the healing pathway transformation.

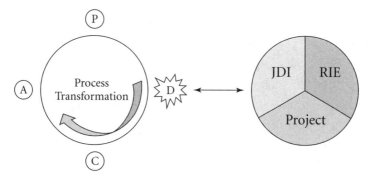

Figure 2.6   The three process transformation tools.

improvement events (sometimes called RIEs or *kaizens*), and projects (Six Sigma, simulations, 5S). Each of these is a process transformation tool, as shown in Figure 2.6.

▲ *Plan* the process transformation. This will be accomplished in various ways depending on whether the process improvement effort is a
  ▼ Rapid improvement event—use a team.
  ▼ Project—use an engineer or small team and Six Sigma or other Lean tools.
▲ *Do* the process transformation.
  ▼ Make the change.
▲ *Check* the process transformation.
  ▼ Provide process-level oversight, including process audits, to ensure the process transformation is achieving sustainable results.
▲ *Act* to eliminate barriers and standardize the process transformation.
  ▼ Standardize new procedures and work instructions.

CHECK AND STUDY

The PDCA cycle is sometimes distinguished from the PDSA cycle. In this text we maintain consistent use of the term PDCA, but intend it to include both Check (ensuring that progress is made and eliminating barriers) and Study (learning and adapting upon review of the results).

CHAPTER 3

Implementing Lean in Healthcare

As discussed in Chapter 2, Lean transformation requires transformational efforts at three cascading levels in the organization: the enterprise level, the healing pathway/value stream level, and the process level. At the enterprise level, the involvement, leadership, and guidance of the senior leadership team are crucial. But it is at the healing pathway/value stream level that the true work of Lean transformation begins. It is the healing pathway/value stream analysis and follow-on activities that yield improvements in process and transformations in culture.

Note: This chapter uses the terms *value stream* and *value stream analysis*. The intent is to create a bridge for those experienced Lean practitioners familiar with these terms. In Chapter 4, the book fully adopts and utilizes the terms *healing pathway* and *healing pathway analysis* to emphasize the role of the patient in healthcare processes.

What Is a Value Stream Analysis?

A *value stream analysis* (VSA) is a structured and facilitated process through which a cross-functional team progresses methodically through a series of interactive and evaluative steps aimed at eliminating waste and improving the flow of patients, products, or services, through the value stream.

The concept of flow is important: *Flow* is an ideal process state in which patients, products, or materials move smoothly through the process with no delays or bottlenecks. Think about patients flowing seamlessly through the emergency department or claims flowing smoothly through the revenue cycle.

> ## HINT
>
> The value stream analysis itself does not involve changes to the value stream. Rather, it yields a rapid improvement plan that will bring about change.

It is important to note that the VSA itself does not involve any changes to the value stream and its processes. Instead, the VSA is a group analysis and planning effort that yields a rapid improvement plan. It is the execution of this rapid improvement plan that creates change.

What Is the Difference Between a Value Stream and a Process?

A *process* is a set of steps that transform one or more inputs into one or more outputs. Processes can often be thought of as comprising four broad components through which the transformation is accomplished: people, procedures, information, and equipment. One simple example of a process is a nurse drawing blood from a patient.

A *value stream* is a set of one or more processes that result in the delivery of a service or product. Value streams are typically high-level, or macro, processes. Value streams are typically composed of multiple processes. For example, an emergency department from the point at which a patient enters until discharged is a value stream, and it is comprised of multiple processes, as shown in Figure 3.1.

The Value Stream Transformation Model

The value stream transformation model is depicted in Figure 3.2. Over the course of the VSA, the team will identify a number of improvement opportunities. Rapid improvement events (RIEs), Just Do Its (JDIs), and projects are the tools by which these improvements will be achieved. The appropriate tool will depend on the type, difficulty, and complexity of the improvement opportunities the team identifies. The team will capture all the opportunities, and assign the appropriate tool, in the rapid improvement plan. This will be the most important team output: a plan for trans-

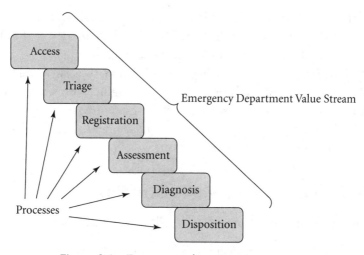

Figure 3.1   Emergency department processes.

forming the current state of a value stream or process to the desired future state.

A rapid improvement plan will typically involve multiple Just Do Its, rapid improvement events, and projects. It will also identify who is responsible for leading each effort, a date by when it will be completed, and a review cycle.

Here is a brief description of each of the improvement tools. You will learn how to do these in subsequent chapters.

▲   **Rapid improvement event (RIE):** This is sometimes referred to as a *kaizen.* A rapid improvement event is typically a 3- to 5-day, facilitated event, aimed at further refining and actually implementing improvements to portions of a value stream.

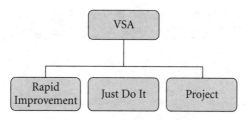

Figure 3.2   Tools for implementing improvements identified in a VSA.

▲ **Just Do It (JDI):** A Just Do It is an action for which no additional study or decision making is necessary. The team has decided it should be done, and all that is left is for someone to make it happen.

▲ **Project:** A project is an improvement effort that typically involves data analysis and is often conducted by an engineer or small team to determine the best approach for achieving a desired outcome. Projects include Six Sigma projects, visual management, mistake proofing, kanban systems, and setup reduction.

How Do I Improve a Value Stream?

Improving a value stream involves the following foundational steps:

▲ Define value from the customer's (patient's) perspective.
▲ Understand the value stream and those steps that are value-added and non-value-added (waste) from the customer's perspective.
▲ Make the value stream flow and reduce the cycle time by driving out waste.
▲ Continually strive for improvement.

At its essence, *Lean* can be summarized as *a process improvement approach that focuses on adding value by identifying and eliminating waste and non-value-added activities from customer-driven value streams.*

Value-added activities are activities performed during the production or delivery of a service or product that increase its value to the customer. To be a value-added action, the action must meet all three of the following criteria:

1. It meets a customer need.
2. It must be done right the first time.
3. The action must change the product or service in some manner.

HINT

At its essence, Lean is about adding value by identifying and eliminating waste and non-value-added activities from customer-driven value streams. Its benefits are improved quality and throughput and reduced cost.

Non-value-added activities are activities performed during the production or delivery of a service or product that utilize time or resources, but do not increase its value to the customer. There are two types of non-value-added activities. The first type is those activities that are non-value-added but required, due to regulatory, legal, or other requirements. The second type is those activities that are non-value added and are not required. This second type is also called waste.

Waste is any type of activity performed during the production or delivery of a service or product that (1) utilizes time or resources, but does not increase value to the customer and (2) is not required for legal, regulatory, or other reasons. There are eight types of waste, and the first letters of each form the mnemonic DOWNTIME, as shown in Table 3.1.

Prior to Lean transformation, a typical value stream includes very little value-added activity. Consider an emergency department with an average length of stay of 4 hours. Most administrators and clinicians, when asked, will readily acknowledge that perhaps 30 minutes of that are value-adding. The remaining 3½ hours are non-value-adding, i.e., waste. Figure 3.3 illustrates this theme. It is likely many of you have had this very experience. From a patient's perspective most of the time spent in an ED is waiting, a common waste.

Table 3.2 provides examples of the eight wastes in healthcare.

Table 3.1   DOWNTIME

Defects	Work that contains errors, rework, or mistakes or lacks something necessary
Overproduction	Producing more than the customer needs right now
Waiting	Idle time created when material, information, people, or equipment is not ready
Not Utilizing People's Abilities	Not using people's mental, creative, and physical abilities
Transportation	Movement of product that does not add value
Inventory	More materials, parts, or products on hand than the customer needs right now
Motion	Movement of people or machines that does not add value
Excess Processing	Effort that adds no value from the customer's viewpoint

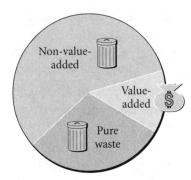

Figure 3.3 Value-added, non-value-added, and waste.

Table 3.2 Examples of the Eight Wastes in Healthcare

Defects	Medication error. Wrong patient. Wrong procedure. Missing information. Redraws. Poor clinical outcomes.
Overproduction	Drawing too many samples. Extra tests.
Waiting	Waiting to see a doctor. Waiting for a procedure. Waiting for a bed. Waiting for testing. Waiting for test results. Waiting for lab results. Waiting for discharge.
Not Utilizing People's Abilities	No empowerment. "Check your brains at the door" mentality. Old guard thinking, politics, or the culture. Poor hiring practices. Low or no investment in training. High turnover.
Transportation	Moving samples. Moving specimens. Moving patients for testing. Moving patients for treatment. Moving equipment.
Inventory	Pharmacy stock. Lab supplies. Samples. Specimens waiting analysis. Paperwork in process.
Motion	Searching for patients. Searching for medications. Searching for charts. Gathering tools. Gathering supplies. Handling paperwork.
Excess Processing	Multiple bed moves. Multiple testing. Excessive paperwork. Unnecessary procedures.

The Value Stream Analysis Event

The term *value stream analysis* refers to the process of analyzing a value stream. Value stream analysis is a powerful tool for analyzing the clinical and business processes within your organization and for identifying opportunities for improvement. This analysis typically occurs during a 2- to 3-day event, with a cross-functional team, and is typically referred to as a *VSA event,* or simply a VSA.

There are five foundational components of any value stream analysis:

▲ Set it up.
▲ Understand the current state.
▲ Envision an ideal state.
▲ Design an achievable future state.
▲ Develop and execute a rapid improvement plan.

Set It Up

This step occurs prior to the event, and it includes the following:

1. Get the commitment of the senior leadership team and the executive with responsibility for the healing pathway.
2. Document a formal charter, including identifying the team members and the team's objectives.
3. Schedule and prepare for the actual event.

Understand the Current State

This is the first portion of the team event. It includes these steps:

1. Document the suppliers – inputs – process – outputs – customers (SIPOC) diagram.
2. Identify the value and how it will be measured.
3. Identify the wastes in the process.
4. Identify the impediments to healing in the process.
5. Document a current state map.

Envision an Ideal State

The word *ideal* suggests that this is not an achievable state, at least in the near long term. The value of this team activity is to generate very creative thinking and ideas. Once the idea state is envisioned, the team is asked to identify the guiding principles that were implicit in the ideal design. These guiding principles are then used to guide the design of the achievable future state.

1. Envision an ideal future state.
2. Capture the guiding principles implicit in that ideal state.

Design an Achievable Future State

Using the ideal state as a starting point, and considering the constraints to their full implementation, design a future state that is achievable in the near term:

1. Design an achievable future state.
2. Prioritize the improvement activities.

Develop and Execute a Rapid Improvement Plan

Document the various opportunities identified for achieving the future state. Some will need no further study and will be identified as Just Do Its. Many of these opportunities will require further study and development and therefore additional work in the form of rapid improvement events or projects. Document these required activities in a rapid improvement plan.

1. Determine the best approach for achieving improvements (JDIs, RIEs, projects).
2. Document these in a rapid improvement plan.

Typically, VSAs are very engaging, exciting, and yes—perhaps exhausting—events. Team members and leaders make a huge commitment of time and energy to make them happen. This commitment is typically rewarded. At the end of the event, leaders and team members alike often say they find these events to be some of the most enlightening and rewarding experiences they have had. In Chapter 4 you will learn how to conduct such a VSA.

Healing Pathway Analysis

What Are Healing Pathways?

Simply put, healing pathways are value streams through which patients flow.

The term *healing pathway* was chosen very intentionally to emphasize a focus on the patient's perception and experience in every improvement activity. In practice, healthcare organizations may continue to use the term *value stream* for business and support processes and the delivery of other products and services. However, adoption of the term *healing pathway* will help engender an appropriate focus on the patient.

Why Make a Distinction Between a Healing Pathway and a Value Stream?

There is one critically important reason for making a distinction between a healing pathway and a value stream: a value stream analysis ignores the perspective of the product or service that flows through it. Its objectives are to improve throughput and increase quality. These objectives are also important for patients. But they are not enough. Patients, unlike widgets,

HEALING PATHWAY

A Value Steam Through Which Patients Flow

Traditional value stream analyses ignore the perspective and individual needs of the product or service flowing through them. In healthcare, the patient's experience needs to be a primary focus.

perceive their environment and respond to it. To facilitate the health and well-being of patients, not to mention their satisfaction, there must be a greater focus on the fact that they are living, perceiving beings who interact with their environment.

Healing pathways are the core of the healthcare industry. Healing pathways, and not value steams, should take primacy.

Value Stream Analyses Focus on Waste and Throughput

Consider a typical value stream analysis. During the process walk, the team looks for bottlenecks to process flow. They capture cycle times, lead times, queue lengths, *first-time quality* (FTQ), and other metrics of importance to the flow of the process. Then, during the waste walk, the team identifies occurrences of the eight wastes with the intent of driving them out of the process. Waste walks are extremely beneficial, perhaps the single greatest tool in the Lean tool kit. Consider the following list of wastes that might be found in an emergency department waste walk:

1. Perception of downtime
   - ▼ Staff working or not working
   - ▼ Five to six people standing and waiting
   - ▼ Staff on Facebook
2. Waiting
   - ▼ Wait of ER physician for hard copy of files
   - ▼ Wait of technician for ED to bring patient
   - ▼ Wait for registration to get labels
3. Undefined processes (ownership)
   - ▼ Radiology ED priority
   - ▼ Calling of MD offices for lab results
   - ▼ Multiple processes for radiology procedures
   - ▼ Wait for radiology results retrieval
4. Patient tracking (bed board)
   - ▼ Manual bed board
5. Documentation of ED patient flow
   - ▼ Organization of charts
   - ▼ Too many radiology order locations
   - ▼ Patient tracking limited

   ▼ Wait for doctors to enter orders

6. Different processes by shifts

   ▼ Delivery of specimens to lab on third shift

   ▼ Specimen pickup not done 24/7

   ▼ Different processes at different times

7. Overproduction

   ▼ Reason for triplicate form at patient registration?

   ▼ Radiology orders to print copies

8. Motion/walking

   ▼ Patient walking to registration from triage

   ▼ Staff in multiple areas

   ▼ Walk from label machine to RN

9. Defects

   ▼ Redraws (1 to 2 per shift)

   ▼ Urine specimen not properly labeled

   ▼ Lack of patient Hx (radiology)

These are all important observations. Eliminating these wastes will improve quality and throughput. But almost all have to do with the process flow, impediments to the flow, workarounds, errors, delays, etc. The elimination of some of these wastes, for example, patient waiting, do improve the patients' experience. But for the most part they do not. The focus is on the process, not the perception of the patient in it.

Elimination of the eight wastes is a necessary, but insufficient step towards improving healthcare. The eight wastes don't capture the entirety of the process. They largely miss the patient experience. A hospital could have a perfectly flowing value stream and yet not be providing a great patient experience!

What Is a Healing Pathway Analysis?

A healing pathway analysis is a value stream analysis that includes evaluation and improvement of the patient's experience. In practice, this involves having the team evaluate the process from the patient's perspective. Additionally, a team member or members should play the role of a patient as they move through the healing pathway.

This evaluation can be done at the same time as the waste walk, but an organization will reap the greatest value by considering it a separate exercise.

HEALING PATHWAY ANALYSIS

The Two Primary Tasks of a Healing Pathway Analysis

▲ Eliminate barriers to process flow
▲ Eliminate impediments to patient healing and satisfaction

The team, and the "patient," will look for the *eight impediments to healing*. These are the equivalent of wastes in a value stream, and they should be eliminated to the greatest degree possible. Just as the eight wastes of a value stream can be remembered by using the mnemonic DOWNTIME, the eight impediments to healing can be remembered by using the mnemonic SICKNESS, as shown in Table 4.1.

Table 4.1 The Eight Impediments to Healing

Stress and anxiety	A common state of patients created by the malady that brought them to the hospital, uncertainty and fear about what may be wrong with them, and unfamiliarity with the hospital setting
Inactivity and waiting	Idle and unproductive time created when staff cannot tend to patients at a rate appropriate to their treatment
Coldness or apathy	An aloofness or distancing from the patient by one of his or her caregivers
Knowledge gap	The information that patients lack about what is wrong with them, what is happening to them, and what is going to happen when
Neglect	The absence of steady interaction and information sharing with the patient
Embarrassment	A negative patient experience caused by a lack of dignity in the treatment process
Submission and helplessness	A state of learned helplessness exacerbated by differentials in information, power, and social status
Statistic	Depersonalization of patients, e.g., "the chest pain in room 13" and "5 boarders in the ED"

Just as eliminating waste from value streams stimulates flow and throughput, eliminating impediments from healing pathways stimulates patient satisfaction and recovery. The tasks associated with a healing pathway analysis are therefore twofold:

▲ Eliminate barriers to process flow
▲ Eliminate impediments to patient healing and satisfaction

A final note is in order: During the course of a rapid improvement event—this is an event during which a team is making changes to improve a healing pathway—the team should consider incorporating the eight enablers of patient healing to the greatest degree possible. These are shown in Table 4.2.

Table 4.2 The Eight Impediments and the Eight Enablers

Stress and anxiety	Calm and comfort
Inactivity and waiting	Progress
Coldness or apathy	Caring and warmth
Knowledge gap	Abundant communication and understanding
Neglect	Engagement
Embarrassment	Dignity
Submission and helplessness	Respect and empowerment
Statistic	Special

CHAPTER 5

Engage the Senior Leadership Team

Why Is It Important?

Transformation of a healing enterprise will not be successful without the full commitment and involvement of the Senior Leadership Team. While senior leadership support is the key to success in any industry, it is even more so in healthcare with its extreme complexity. Complete alignment of the senior leadership team behind a well-designed and well-managed transformation plan is critical to its success.

▲ **Commitment:** The commitment of the CEO and Senior Leadership Team must be sincere, visible, and based on a foundation of understanding the return on investment (ROI).

▲ **Involvement:** The CEO and Senior Leadership Team must be actively and visibly involved in leading the transformation.

How Do I Do It?

Step 1: Make the Business Case

The key to getting the commitment of the CEO and senior leadership team is to educate them about the benefits to be achieved. These come in at least three different forms:

▲ Improved patient safety, care, and satisfaction
▲ Improved team member and physician satisfaction
▲ Improved operational performance

LEADERSHIP TEAM
Suggested Messaging for Leadership Team
Lean is a proven approach for creating customer value. It focuses on the identification and elimination of waste and non-value-added activities from healing pathways and value streams. The result is improved patient care and satisfaction, increased team member and physician satisfaction, and improved operational performance. In addition to being a tool for process improvement, Lean helps bring about a cultural transformation. During Lean improvement events, team members are empowered to objectively evaluate their work processes and to create and achieve improvement objectives. This empowerment and involvement stimulates cultural change from the bottom up.
Ultimately, Lean is a tool that the organization's leaders can draw upon. Performance improvement experts, such as Lean black belts, facilitate events, provide education on Lean concepts and their application, and offer recommendations for process improvement. However, process owners and their teams are the only ones who can actually make, and hold, the gains. They own the responsibility for the system, and they deserve the credit for system improvements. Healing enterprise leaders have an opportunity to embrace the Lean philosophy and take advantage of the benefits it offers.

There are numerous sources of information that can be used to make the business case. An initial set of resources is provided in Table 5.1. You can use these or develop your own white paper or presentation for the leadership team.

Step 2: Provide Basic Education and Training on the Process and Their Responsibilities in It

Once you have made the business case, you will want to provide basic education and training to the senior leadership team. Ideally, this session would be conducted as a two-day off-site event. These are the suggested objectives for this initial executive leadership training:

Table 5.1 Recommended Resources

Chapter 1 of this book.
M. Dean, *Healing Healthcare: A Leadership Journey*, GOAL/QPC, Salem, NH 2012.
"Going Lean in Healthcare," Diane Miller, editor. *Going Lean in Health Care*, Institute for Healthcare Improvement, Cambridge, 2005.
Louis Savary and Clare Crawford-Mason, *The Nun and the Bureaucrat: How They Found an Unlikely Cure for America's Sick Hospitals*, CC-M Productions Inc., Washington DC, 2006.
S. Welch, *Quality Matters: Solutions for a Safe and Efficient Emergency Department*, Joint Commission on the Accreditation of Healthcare Organizations, Oakbrook Terrace, Illinois, 2009.

▲ Gain a basic understanding of Lean Sigma principles and tools.
   ▼ Healing pathway/value stream analyses
   ▼ Rapid improvement events (kaizens)
▲ Understand the Lean Sigma transformation process.
▲ Understand leader's roles in the transformation.
▲ Have fun!

A NOTE OF CAUTION

Many of us have been part of transformation efforts that failed. Often we build our employees' expectations, only to have the organization's commitment wane.

If you are not absolutely certain that you have the organizational will to adopt and complete a full transformation, you are better off doing what you can, one healing pathway/value stream at a time.

In such a case you would be well advised to skip the chapters that focus on developing the infrastructure for a complete transformation, and instead select one key process for transformation. Begin there with a healing pathway/value stream analysis; and if you are successful, then you may begin to generate the energy and excitement needed for transformation.

Keeping in mind the nuances of adult, and in this case executive, education, you will want to ensure that there is a balance of didactic information and experiential exercises. Table 5.2 provides a suggested agenda for a 2-day executive leadership session.

Table 5.2 Suggested Agenda for a Two-Day Executive Leadership Session

Day 1	
0730 – 0800	Meet and Greet: Continental Breakfast
0800 – 0900	Lean Enterprise Overview
0900 – 1200	Lean Enterprise Simulation
1200 – 1300	Lunch
1300 – 1430	Waste and Suboptimization
1430 – 1645	Lean Tools Part I
	5S
	Error Proofing
	Visual Controls
1645 – 1700	Wrap Up and Evaluations
Day 2	
0730 – 0800	Meet and Greet: Continental Breakfast
0800 – 1200	Value Stream Analysis
1200 – 1300	Lunch
1300 – 1400	Theory of Constraints
1400 – 1430	Rapid Improvement Events
1430 – 1630	Lean Tools Part II
	Histograms
	Paretos
	The Normal Curve
	Statistical Process Control
1630 – 1730	Effective Sponsoring
	Wrap Up and Evaluations

Step 3: Reinforce the Criticality of Their Participation and Get Their Commitment to the Following Responsibilities

CEO

▲ Champions change and improvement.

▲ Energizes transformation of the healing enterprise.

▲ Sets expectations and holds leaders accountable.

▲ Creates an environment conducive to improvement and change.

▲ Removes barriers to success.

Senior Leadership Team

▲ Sets expectations for middle management concerning performance improvement and holds managers accountable.

▲ Leads interdepartmental initiatives to achieve goals.

▲ Monitors progress toward those goals.

▲ Charters and monitors the progress of key functional and cross-functional improvement projects.

▲ Removes barriers to success.

VP of Performance Improvement

▲ Provides staff leadership to the healing enterprise transformation.

▲ Sits on the senior leadership team and works with senior leadership to translate strategic and annual business plans into performance improvement projects.

▲ Provides technical data analysis, process improvement, and change in leadership expertise, working with leaders and managers to identify, implement, and monitor the success of improvement ideas.

▲ Facilitates key strategic process improvement projects.

▲ Supervises other process improvement specialists (either part-time or full-time resources).

▲ Reports to the CEO, COO, or CFO to increase face validity and maintain the objectivity to work with all areas of the organization, as appropriate.

All

▲ Each of the senior leaders will also provide leadership as members of the transformation leadership team and as executive sponsors at the process transformation level. These responsibilities are described in Chapter 6.

PART II

Healing Enterprise Transformation

Develop and Document the Governance Structure

What Is It and Why Is It Important?

The governance structure is the enterprise's hierarchy for leading and executing the transformation. It includes the group of executives who will lead the organizational transformation. It also includes the organizational hierarchy of leadership and transformation teams, their respective authorities and responsibilities, and the schedule of meetings and events that will maintain the drumbeat of the transformation.

The governance structure is critically important to the success of your organizational transformation. It both defines the required involvement of the key leaders and sets expectations for participation, leadership, and constancy of purpose. It also provides the process for ensuring consistent and regular follow-up to the organizational transformation process.

Once the governance structure is defined, you will want to document your teams, their roles and responsibilities, and the expected level of effort in a transformation governance document.

How Does This Fit into the Big Picture?

Remember the enterprise transformation process from Chapter 2? It is reproduced here as Figure 6.1.

The enterprise transformation cycle comprises the executive-level process for planning and executing the overall transformation. It is a function of the senior leadership team. The healing pathway transformation

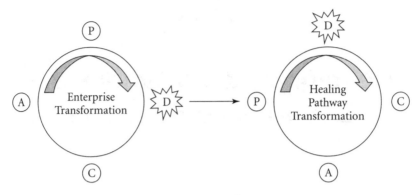

Figure 6.1 Enterprise transformation process.

cycle comprises the planning, execution, and standardization of improvements to operational processes.

Both of these cycles require organizational infrastructure to help them thrive. This chapter describes the governance process for executing both these cycles and ensuring transformation of your healing enterprise.

What Do I Do?

▲ Define the transformation hierarchy.
▲ Determine leadership roles and identify the individuals by name.
▲ Develop team and individual job descriptions for the transformation.
▲ Develop a team meeting and review schedule.
▲ Document all in your transformation governance document.
▲ Get the approval of the CEO and senior leadership team (SLT).

How Do I Do It?

Step 1: Define the Transformation Hierarchy

There may be no single model that is both necessary and sufficient for ensuring success. However, any successful model will include the following elements:

▲ It will be led by the CEO or his or her designee with full, absolute delegated authority.

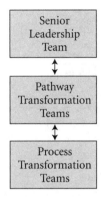

Figure 6.2   Transformation organizational chart.

▲ It will include the key support executives (CFO, CIO) and operational executives (COO, hospital CEOs, CNOs, CMOs).

▲ It will rely on advice from both a physician advisory committee and a committee of the trustees.

▲ Use the transformation organization chart depicted in Figure 6.2 as a model, but make it your own.

Step 2: Determine Leadership Roles and Identify Individuals by Name

You will want clear lines of responsibility and authority for a successful transformation. The CEO will most likely be the de facto head of the senior leadership team (SLT). The vice presidents (VPs) over the respective process areas, for example, surgical services, revenue cycle, and supply chain, will be

CASE STUDY

Healing University Healthcare System decided to pursue organizational transformation in conjunction with its move to a new core information technology (IT) system. Consequently, its governance structure included components from IT, Policy and Procedure, Communication, and Education in addition to the process transformation teams. The governance structure is depicted in Figure 6.3.

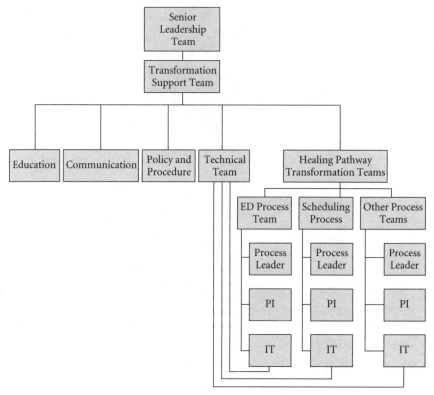

Figure 6.3 Healing University Healthcare System transformation governance structure.

the most likely candidates to lead the healing pathway transformation. What remains is to identify the transformation executive. This should be a senior VP or well-respected VP, for example, the VP of performance improvement.

Transformation Support Team In some organizations it may be practical for the senior leadership team to delegate responsibility for the transformation to a subset of the senior leadership team, called the transformation support team. The CFO or COO might be an appropriate choice to act as executive sponsor for this, with the VP for performance improvement acting in the role of team lead. It will depend on the authority and respect these various positions command within your organization. The bottom line is that either the SLT or the transformation support team needs to be the driving and coordinating force behind the transformation,

> ## HINT
>
> You will review and finalize the governance structure and team member-ship at the transformation summit. The work you do now will both serve as a guide and move you farther down the path of transformation!

and you will want that team to have the organizational authority to make things happen.

These assignments should be finalized to the degree possible prior to the transformation summit (described in Chapter 7). However, if they are not finalized prior to the transformation summit, these assignments can be discussed and finalized by group consensus during the transformation summit. Both key leadership positions and team members will be identified during the transformation summit.

Step 3: Develop Job Descriptions for the Transformation Teams

To achieve their buy-in and commitment, you will want the leaders of the organization to participate in determining the roles and responsibilities of the various teams that will participate in the transformation. You may wish to start with the following templates and finalize them through a group decision-making process during the transformation summit.

Senior Leadership Team

Members: CEO; all senior vice presidents; the transformation executive; a physician leader

Activities

▲ The SLT sets expectations for pathway transformations regarding performance improvement and holds executive sponsors and their teams accountable.

▲ The SLT leads interdepartmental initiatives to achieve goals.

▲ It monitors progress toward those goals.

▲ It ensures the availability of resources.

▲ The SLT removes barriers to progress.

Transformation Support Team This team is optional as it is not needed if the SLT provides this leadership.

Members: CFO; COO; VP or senior vice president (SVP) of performance improvement; hospital CEOs; finance; IT; nursing; physician leader

Activities

▲ The transformation support team reports to the senior leadership team.

▲ The organization COO or CFO may serve as the executive champion for the group, and the VP of performance improvement may serve as the team leader.

▲ The transformation support team charters teams and approves memberships.

▲ It approves resource needs.

▲ It approves priorities and time lines.

▲ It addresses obstacles and removes barriers.

▲ The transformation support team defines the accountability of teams.

▲ It facilitates system communication.

Healing Pathway Transformation Teams

Members: VP or SVP to act as executive sponsor; senior leader over the process to be transformed will be team leader; black belt; subject matter expert; IT expert for systems involved with the process; a cross-functional representation of team members intimately involved in the process.

Executive Sponsor and Team Leaders Transformation Teams

Executive Sponsors

▲ These executives have authority to approve or disapprove process redesign changes.

▲ They are accountable for team progress and barrier removal.

Team Leaders

▲ Typically, the team leader is the process owner.

▲ Team leaders are responsible for moving the team through its agendas and plans.

▲ They are responsible for team logistics.

▲ They are the key interface with the sponsor.

Activities

▲ Process teams are charged with identifying and implementing Best Practices for their assigned process.

▲ Process teams monitor and report results.

▲ They achieve core measures or benchmark metrics.

▲ They set service expectations.

▲ They standardize policies and procedures.

▲ They report results and sources.

▲ Process teams seek assistance from the executive sponsor and/or transformation team as needed.

Education Team

Members: VP/director of corporate training; members of process teams; IT experts; other individuals with specific expertise as needed.

Activities

▲ The education team supports process teams by providing education and training on new processes to team members not involved in the redesign.

▲ It provides training and education on new IT systems and solutions as needed.

▲ It may provide broader training in Lean Sigma improvement techniques and methods.

Communication Team

Members: VP of communications; transformation executive; other interested parties.

Activities

▲ The communication team brands transformation efforts across the organization.

▲ It creates ongoing communication tools (e-mail newsletter) at least monthly.

▲ The communication team gets out the quarterly report from the process teams.

Policy and Procedure Team

Members: VP or director of compliance or commensurate position; representatives of active process teams.

Activities

▲ The policy and procedure team ensures configuration control of already developed policies and procedures.

▲ It documents new procedures to assist in standardizing Best Practices.

▲ It liaises with process teams and accrediting bodies.

Technical Team

Members: CIO; IT system experts

Activities

▲ The technical team supports process teams by making appropriate changes to IT system-driven processes.

▲ It ensures systemwide compatibility of all IT changes.

▲ It assists in the implementation of Best Practices.

Step 4: Develop a Team Meeting and Review Schedule

All team activities should be coordinated to ensure a constant drumbeat of transformation. This includes the schedule of all team meetings from that of the senior leadership team to each of the process teams. The periodicity of such meetings may vary over time. Typically, you will want the drumbeat to be more frequent near the beginning and middle of the transformation, fading a bit as the core of the transformation comes to a successful close.

As with the other governance infrastructure and process steps, you will want to discuss and finalize these at the transformation summit. But as a starting point, you may wish to consider the following meeting intervals.

Senior Leadership Team. The meeting interval should be biweekly initially, fading to monthly after the first 8 to 12 weeks of successful transformation efforts.

Transformation Support Team. The meeting interval should be weekly until very late in the transformation process, fading to monthly as the role changes to one of identifying opportunities for continual improvement rather than leading the transformation.

Healing Pathway Transformation Teams.   Meetings should be held weekly or biweekly after the initial healing pathway/value stream analysis. These meetings come to an end once the new process is documented, all IT and other changes have been implemented, and the new process is fully hardwired.

Education Team.   Meetings should be held weekly as the needs of the process teams are assessed as appropriate training and development is designed and provided.

Communication Team.   Meetings should be held biweekly initially, fading to monthly and then quarterly as the transformation progresses.

Policy and Procedure Team.   Meetings should be held weekly. This team will be active as long as process teams are underway and then for a few months thereafter until all process changes are documented and hardwired.

Technical Team.   Meetings should be held weekly. As with the policy and procedure team, this team will be active as long as process teams are underway and then for a few months thereafter until all process changes are documented and hardwired.

Step 5: Document the Governance Structure in Your Transformation Governance Document

Capturing all the work you have done to this point will help ensure your success by memorializing it and allowing all participants to have a record— or instruction guide if you will—of their commitment. This allows for review, modification, and ultimately buy-in and commitment of those who participate in its development.

Step 6: Get the Approval of the CEO and SLT

If you followed the process as outlined above, your CEO and the SLT should have been very involved in the development of the governance structure and process. If not, at a minimum you will need their review of and commitment to the documented plan.

HINT
Developing the governance structure is the key deliverable here. But don't underestimate the importance of documenting it in the transformation governance document.

CHAPTER 7

Plan the Lean Transformation Summit

What Is It and Why Is It Important?

The Lean transformation summit is the signature event for initiating the transformation of your healing enterprise. Simply put, it is a multiple-day planning session during which the senior leaders of an organization:

▲ Achieve a shared vision of what the transformed organization will look like and what they want from it

▲ Gain consensus on guiding principles (how they will work together as well as how the process will work)

▲ Develop and document their enterprise map to understand the highest-level processes of their enterprise

▲ Identify and prioritize these high-level processes for transformation

▲ Develop a transformation master schedule

▲ Review, validate, and/or complete the transformation governance process (and document) to include all leadership roles and responsibilities

▲ Identify and/or confirm the leadership and membership of all process transformation teams and officially charter them

The Lean transformation summit is important for the following reasons:

▲ It will demonstrate to all the commitment of the CEO and senior leadership team to the transformation process.

▲ As a participative planning session, it will engender a shared commitment to the transformation process.

▲ It will yield a clear road map for transformation.

This will be a huge event for the organization—the real first step in the transformation of your healing enterprise. You want it to go smoothly. Detailed planning is the key to making that happen. This planning will include frequent dialogue with the CEO or executive sponsor and careful visualization and orchestration of how the events will unfold over the course of the summit. You may wish to use consultants to help you plan and perhaps facilitate the summit.

What Do I Do?

▲ Develop an agenda.

▲ Coordinate with the CEO to kick it off.

▲ Determine a location and schedule.

▲ Have an invitation sent out from the CEO.

▲ Plan the interactive events and ensure you have the right supplies.

▲ Determine if you want to use an outside facilitator.

How Do I Do It?

Step 1: Develop an Agenda

An agenda should always flow from objectives. Carefully consider what your objectives for the summit are. In other words, what outcomes do you hope to achieve? These can be both tangible and intangible. Tangible outcomes include identification of processes to transform, schedules, assignments, etc. Intangible outcomes include shared commitment, stronger interpersonal relationships, etc. You probably want some of both.

The "What Is It and Why Is It Important?" section of this chapter provides a list of objectives. You may wish to review that list as a starting point, adding to or subtracting from it to meet your unique situation. Then you will want to adapt the sample agenda provided later in this chapter accordingly.

Step 2: Coordinate with the CEO to Kick It Off

You will want the CEO to model commitment to the transformation by kicking off the transformation summit. You will also want to keep her or him fully informed during the planning stage and solicit any input that might be offered. You will want your CEO to make the message his or her own, but the message would do well to elaborate on the following points:

▲ Give a brief "state of the union" summary about how we are currently doing.

▲ Offer a reminder of the mission and vision of the organization.

▲ Talk with excitement about the benefits of Lean in facilitating organizational change and transformation.

▲ Discuss external factors and what is happening nationally.

▲ Transformation is one of the most important things we will do over the next three years.

▲ Our people, including this team, have a lot on their plates. We need to prioritize our work so that we

▼ Address today's challenges.

▼ Build a foundation for tomorrow.

▲ You are the leaders of this healthcare system. We need to increase your participation in making our vision a reality. This summit is intended to be the starting point for how we do that.

Step 3: Determine a Location and Schedule

While it is sometimes difficult to get people away from the office, an off-site event is the best option for getting, and keeping, the team fully engaged in planning the transformation. You may wish to consider a local venue to make it convenient. The event will take 2 to 3 days, and it is best to do it at one time. However, doing a day at a time, with a week in between, is also effective and may be more amenable to people's schedules. You will want to publish the schedule 4 weeks or more in advance to ensure everyone's ability to participate.

With regard to the facilities themselves, a venue with a large conference area will be most convenient. You will want your team to be comfortably seated at round tables, each with a flip chart for team exercises and reports. You may also wish to have break-out rooms available depending on the acoustics of the large room. At some points during the summit you will have multiple teams energetically engaged in their respective efforts.

You will also want to arrange to have breakfast and lunch catered in. You will want the team to stay in place and engaged over the course of each day.

Step 4: Have an Invitation Sent Out by the CEO

This is another sign of the CEO's commitment, and such an invitation will very effectively ensure participation. Figure 7.1 provides a sample letter from the CEO of Healing University Healthcare Systems. Adapt it to suit your own needs!

Step 5: Plan the Interactive Events and Ensure You Have the Right Supplies

The agenda is the starting point and the framework for your summit. But now you need to determine the specific exercises and activities you will use. For example, what process will you use for reaching consensus on the guiding principles? Will you use a nominal group technique and voting with the entire team? Will you break them into groups and then have a report out?

Similarly, how will you identify and prioritize the processes for transformation? Much of this summit will be very interactive, and you will want to carefully plan the process. Once you have determined the appropriate processes to employ, determine the supplies you will need to have on hand and ensure that they are available. At a minimum you will need the following:

▲ Flip chart easels for each table and a couple for the front of the room
▲ Flip chart pads (Bring extras!)
▲ Masking tape
▲ Flip chart markers
▲ Yellow stickies
▲ Markers for writing on yellow stickies
▲ A camera for capturing the teams' work
▲ Projector and screen for your presentation
▲ Stress relievers such as squeeze balls
▲ Refreshments

From the desk of the CEO

Dear Healing University Healthcare Leader:

Because of the work and dedication you and your teams provide every day, we are making great strides in our journey to become the best health system in America. I have recently shared with you the next critical step in this journey: Lean transformation of our healing enterprise. This inititative is going to be critical in helping us accomplish our core objectives and fulfill our vision.

The first significant step is your participation in the upcoming Transformation Summit to be held on October 2 and November 7. The October summit will be held from 8:30 am to 4:30 pm in the Derby room of the Downtown Convention Center, with a reception to follow. The November summit will be held from 8:00 am to 5:00 pm at the same location.

Great care has been taken in designing this summit to engage our top leadership talent. Our objectives are to craft a shared vision of what our organization will look like and a roadmap for getting us there. Expected outcomes include:

▲ A shared vision of what a transformed organization will look like and what we want from it.
▲ Consensus on guiding principles (how we will work together as a team to accomplish the transformation)
▲ Prioritization of our key processes for transformation
▲ A transformation roadmap and master schedule
▲ A governance process for the transformation to include all leadership roles and responsiblities

Please accept this personal invitation with a challenge for you to become fully engaged in this process. As a senior leader, you as an individual, and we as a collective team, are responsible for the future of our organization. This is the way we will get there, and we can't do it without you!

A full agenda and complete workshop materials will be forthcoming. As always, thank you for your leadership.

Sincerely,

Will Patterson
President and CEO
Healing University Healthcare

Figure 7.1   Sample CEO invitation.

Step 6: Determine If You Want to Use an Outside Facilitator

This is a major and significant event. It includes the most senior leaders of your organization, and it needs to provide a smooth process to help them design your future. The design and leadership of such an event can be intimidating to even the most seasoned facilitator. Careful consideration should be given to the following:

▲ Your experience in designing these types of workshops
▲ The availability of a seasoned internal facilitator
▲ The dynamics of the senior leadership team, and specifically their willingness to be led by an internal facilitator

If successful, an event designed and facilitated by an internal leader may well cement that leader's credibility and role going forward. But the converse is also true. Err on the side of using outside help, for both the design of the session and its facilitation.

CHAPTER 8

Conduct the Transformation Summit

What Do I Do?

Follow the plan! The planning you have done as advised in Chapter 7 has prepared you well for this moment. The CEO is committed and ready, and the exercises are planned. You know where you are going, and you know how you are going to get there. This is your moment to shine as well as the really exciting beginning of your organization's transformation!

Your agenda probably looks something like this. Now just do it!

▲ CEO leads kickoff.
▲ Develop a shared vision.
▲ Develop a set of guiding principles.
▲ Develop the enterprise map.
▲ Prioritize the processes to be transformed.
▲ Develop the transformation master schedule.
▲ Identify the executive sponsor and team members for each process.

HINT
You will be well served to visit the location of the off-site event the night before, to ensure everything is just as you want it. Double-check your supplies, and take them to the off-site location the night before the event as well.

> ## HINT
>
> Everyone who participates in the transformation summit should already have been trained in Lean, typically by participation in at least a two-day executive session.
>
> The transformation summit is not intended to be a training experience, but rather the engagement of trained people in planning the transformation.

▲ Update the governance infrastructure and process.
▲ Develop a communication plan.

How Do I Conduct the Transformation Summit?

Step 1: CEO Leads the Kickoff

As described in Chapter 7 on transformation summit planning, the CEO will kick off the transformation summit. In doing so, she or he will model commitment to the transformation, express excitement at this way forward, and provide a call to action for the leadership team. The leader empowered to lead the transformation effort will likely introduce the CEO, framing up the benefits of a Lean transformation and the benefits expected for your organization. You are still in the process of educating the senior leadership team, and this is an excellent opportunity to do so, reinforced by the implicit and explicit support of the CEO. Take advantage of the opportunity.

Step 2: Develop a Shared Vision

Many organizations have developed a mission, vision, and values. Often, the mission statement describes why an organization is in business,

> ## HINT
>
> The *Lean Six Sigma Pocket Toolbook* is an excellent source of guidance for many of the team exercises you will employ during the transformation summit.

CASE STUDY

Healing University Healthcare System

Will Patterson, CEO of Healing University Healthcare System, wanted to instill in his team the notion that the outcome of the transformation summit, and more importantly the success of Healing University Healthcare System, was critically dependent on the interaction of the leadership team. He used the following quote from Peter Senge, and a short reflection that went along with it, to stimulate that thinking.

Every organization is a product of how its members think and interact.
—Peter Senge

discussing the products or services delivered. It is of and for the customers. The vision is often aspirational, and it paints a picture of what the organization might look like if it were performing its mission in an excellent manner. It is of and for the leaders of the organization. The values are of and for all members of the organization, describing what they hold dear.

What you will develop at the transformation summit is a shared vision of what a transformed organization will look like. It should flow from, and be in alignment with, your organizational mission, vision, and values. But it will be more focused on operational elements associated with the transformation. Developing an affinity diagram would be an effective process for generating descriptors for your shared vision of a transformed organization. Depending on the number of participants, you may want to use multiple small groups and then consolidate and use an affinity diagram to organize their top ideas.

You may choose to facilitate an unconstrained visioning session, in which the participants start with a blank sheet of paper and there are no additional guidelines. Conversely, you may wish to focus the team members a little bit more in their work. For example, you could begin with an exercise to identify the key stakeholders and then break into teams and develop visions for each of those.

CASE STUDY

Healing University Healthcare System's Visioning Session

During Healing University's transformation summit, the visioning was conducted around key stakeholders who were identified. The group was divided into teams, each developing a vision around their assigned stakeholder. Then these were reviewed by the larger group and consolidated into a unified vision. Here is the list of stakeholders they identified around which to do their visioning work:

▲ Patients
▲ Administration
▲ Payers
▲ Physicians
▲ Nurses/clinicians
▲ Suppliers

Figure 8.1 provides an example of what one of the Healing University teams came up with for their vision of what they wanted for one of their stakeholders, in this case the patient.

As each team completed its work, key elements and consistent themes were identified. These were integrated into a model depicting a vision of a transformed Healing University Healthcare System, as depicted in Figure 8.2.

Step 3: Develop a Set of Guiding Principles

The guiding principles for the transformation should be in alignment with, but probably different from, your healing enterprise's values. As with the transformation vision, you will want to identify the guiding principles for how you and your team will engage in the transformation, i.e., how you will relate to and interact with one another through the transformation process and what you are trying to achieve at a very high level. Small group work and affinity diagrams may be your tool of choice here as well.

Patient

▲ Healing University Healthcare System is organized. "They were expecting me."

▲ There is a consistent service standard.

▲ Patient-centered care team works together to give seamless care.

▲ There is easy access.

▲ They know me here.

▲ My needs will be met.

▲ They care about my family.

▲ They listen to me.

▲ I feel safe.

▲ They have a technologically competent and compassionate team.

▲ I trust Healing University hospital.

▲ They do the right thing at the right time.

▲ They are accurate, compassionate, great.

▲ They give the best care.

▲ They anticipate my needs.

Figure 8.1   Healing University's vision for its patients.

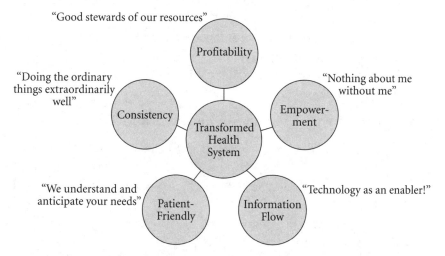

Figure 8.2   Healing University's vision of a transformed Healing University.

CASE STUDY

Healing University Healthcare System's Guiding Principles

During its transformation summit, Healing University Healthcare System concluded that the ultimate objective of the transformation was to "Standardize Best Practices" across the healthcare enterprise. That meant identifying Best Practices, redesigning their processes to achieve Best Practices, and ensuring that the new processes were repeated consistently and reliably. They coined the phrase "the Healing Way" to describe the standardized way they would do things.

Here are the refined results of their "guiding principles" affinity.

Process Transformation: Operational Excellence

▲ The Healing Way: Best clinical and business practices identified
▲ The Healing Way documented in policies and procedures
▲ The Healing Way implemented consistently across the system
▲ Consistent adherence to the Healing Way by all physicians and staff

People Transformation: Distinctive Culture and Leveraged Partnerships

▲ Best leadership practices developed and implemented across the system
▲ Culture of performing in accordance with the Healing Way
▲ Culture of continual improvement
▲ Engaged and empowered team members
▲ Aligned physician partners

Technology Transformation: An Enabler, Not a Barrier

▲ Core Health Information System enables effective and efficient:
  ▼ Patient care
  ▼ Patient flow through all parts of the system
  ▼ Medical records
  ▼ Revenue cycle processes
  ▼ Business information

Step 4: Develop the Enterprise Map

An enterprise map is a high-level diagram of the internal services and activities of an organization. When applied to a healing enterprise, it will identify your core processes for standardization and redesign including both patient care (long-term care, emergency department, patient flow, imaging, diagnostics, surgery, etc.) and revenue cycle (patient accounting, health information management, etc.) and will depict their interdependencies. Developing this map is a necessary first step before prioritizing value streams for transformation. The enterprise map will:

▲ Readily identify the scope of the service processes and associated functions which, if improved, can provide the biggest benefit for the effort expended.

▲ Help the team understand the extent to which services interact with various parts of the organization and the supply chain.

▲ Help teams more accurately define the value of the service and its associated information.

▲ Enable teams to organize all the major resources—people, IT, equipment, policies and procedures, etc.—according to their use in the natural flow of service activities.

▲ Highlight gaps, duplications, and misunderstandings between different functional areas.

▲ Improve team decision making by building a shared understanding and acceptance of current practices, while imagining a desired future state.

▲ Provide the basis for detailed healing pathway/value stream mapping.

Figure 8.3 depicts the high-level enterprise map prepared by the Healing University Healthcare System.

From the view at the enterprise level, you will want to engage the team in brainstorming and identifying the key processes you perform. At Healing University, the team identified the processes in the following case study.

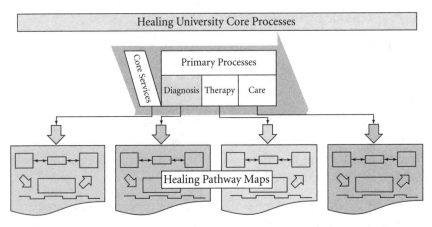

Figure 8.3 Healing University Healthcare System's enterprise map.

CASE STUDY

Healing University's List of Processes to Transform

▲ Revenue cycle front end: Central scheduling/registration/ transcription
▲ Revenue cycle back end: Patient accounting/HIM/reimbursement
▲ Case management
▲ Surgical services
▲ Nursing electronic medication administration record (EMAR)
▲ Emergency department
▲ Supply chain
▲ Medical imaging
▲ Electronic medical record (EMR) clinical data repository
▲ Physician systems (CPOE, EBM)
▲ Bed control
▲ Financial decision support
▲ Physician recruit/relations
▲ Pharmacy
▲ Laboratory
▲ Patient safety
▲ Ambulatory physician practice management

CASE STUDY (*continued*)

▲ Budget and forecasting/corporate accounting
▲ HR (workforce, benefits, education)
▲ Quality/accreditation
▲ IT (help desk, project management)
▲ Infection control
▲ Psychiatry/behavioral health
▲ Dietary/nutrition
▲ Contract management
▲ Home health
▲ Environmental services
▲ Engineering
▲ Security systems
▲ Patient transport
▲ Rehabilitation
▲ Facility/property management
▲ Risk management

Step 5: Prioritize the Processes to Be Transformed

Now that you have identified your key clinical and business processes, you will engage the team in prioritizing their transformation. You will want to give consideration to:

▲ The impact on patient care
▲ The impact on throughput
▲ The impact on profitability
▲ Ease of transformation
▲ Availability of a strong leader

You may choose to give particular consideration to those processes that improve performance now while also building a pathway to the future.

Process ranking and prioritization can be done using any number of team voting techniques. These range from a more complicated rational economic model in which all the decision criteria for evaluating each process are identified and weighted, to a much simpler team-based 3-2-1

HINT

You may wish to prioritize for transformation those processes that improve performance **now** while building a pathway to the future.

---

voting scheme in which each member casts only three votes, allotting three points to the most important criteria, two points to the next most important, and 1 point to the next most important. Do what seems best for your organization based on the interaction of the group, its size, and the characteristics of the leaders and team members you are working with. Healing University Healthcare System came up with the prioritized ranking of processes depicted in Table 8.1.

A Pareto chart of the rankings provides a nice visual of the top eight processes to be transformed. Healing University's results yielded the Pareto chart shown in Figure 8.4.

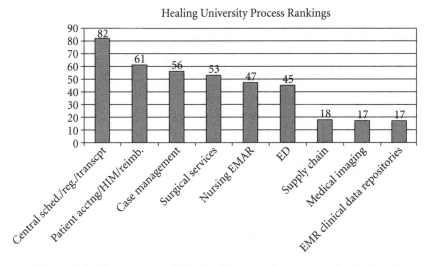

Figure 8.4 Pareto chart of Healing University's process prioritization.

Table 8.1   Results of Healing University's Process Prioritization

Revenue cycle front end: Central scheduling/registration/transcription	82
Revenue cycle back end: Patient accounting/HIM/reimbursement	61
Case management	56
Surgical services	53
Nursing EMAR	47
Emergency department	45
Supply chain	18
Medical imaging	17
EMR clinical data repository	17
Physician systems (CPOE, EBM)	15
Bed control	14
Financial decision support	10
Physician recruit/relations	7
Pharmacy	4
Laboratory	3
Patient safety	3
Ambulatory physician practice management	3
Budget and forecasting/corporate accounting	2
HR (workforce, benefits, education)	2
Quality/accreditation	1
IT (help desk, project management)	1
Infection control	1
Psychiatry/behavioral health	1
Dietary/nutrition	0
Contract management	0
Home health	0
Environmental services	0
Engineering	0
Security systems	0
Patient transport	0
Rehabilitation	0
Facility/property management	0
Risk management	0

Step 6: Develop the Transformation Master Schedule

Now that your team has determined which processes are most important to your transformation, you will want to develop a master schedule. The schedule will be dependent on your priorities, the availability of the appropriate leaders, the availability of team members, and the availability of black belt facilitators and subject matter experts. Many of these resources will be used for more than one process, so it is best to develop first the schedule and then the team memberships to support the schedule. The following case study and Figure 8.5 provides Healing University's transformation master schedule.

CASE STUDY

Healing University Healthcare System developed the following schedule to meet its needs. Note that it is about a 2-year effort, which is about typical for this sort of work. They have time-phased transformation of their prioritized projects based on importance and resource availability.

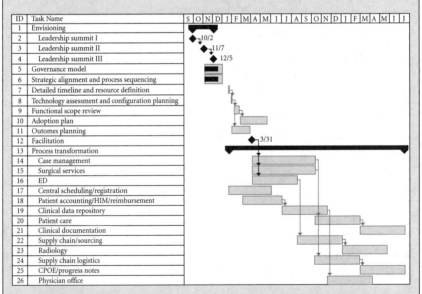

Figure 8.5 Healing University's transformation master schedule.

Step 7: Identify the Executive Sponsor and Team Members for Each Process

You will typically want the senior executive over a given area to serve in the role of executive sponsor. A senior manager under the executive sponsor may serve as the team leader, who will be responsible for the day-to-day leadership of the process transformation. You will recall that these roles and responsibilities have been described elsewhere in this book, and they should also have been part of the executive training that every member of this team has received. Thus that will not be repeated here.

What you will want to do at the transformation summit is draft a charter for each of the teams you have scheduled. A typical charter should provide clarity around the following elements:

▲ Name of the healing pathway/value stream to be transformed
▲ Purpose of the project
▲ Business case
▲ Flow boundaries (where the process begins and ends)
▲ Constraints (what constraints the team should be aware of, i.e., funding, FTE, capital)
▲ Team membership

Figure 8.6 is an example of one of the charters developed at Healing University's transformation summit.

You will probably split the transformation summit participants into teams to develop these charters. Once the teams have drafted their charters, you should conduct an out-briefing of each for the entire summit group. Charters, and especially charter membership, may need to be adjusted depending on the team members desired or required for respective processes.

Step 8: Update the Governance Infrastructure and Process

In Chapter 6, you and your team developed the governance infrastructure and process. Now is the time to review what is there and solicit feedback. In addition, now that you have identified the processes to be transformed and have identified executive sponsors, team leaders, and team members, you will be able to update that information. More importantly, you will be able

Healing University Emergency Department

EVENT TYPE	X HPA	RIE	Project	Just Do It

Healing Pathway: Healing University Emergency Department

Event Purpose: Analyze and understand the current state. Design a desired future state and document a rapid improvement plan to get us there.

Business Impact: The ED is a front door to our system. Improving ED flow will result in reduced LWBS and LOS. Increased patient and team member satisfaction and increased OP and IP volume and revenue.

Flow Boundaries: Start point: Patient walks in the door   End point: Admit or discharge

Constraints: Any capital or other requests must be approved through normal channels. What can we do now with what we have? FTE not to exceed productivity targets.

Data: Data to collect beforehand: LWBS, hours on diversion, LOS, volume, percent ip/op, patients transferred to downtown, cover margin

Team Members: (Name and Department)

Executive Champion: Michael C   Black Belt: Floyd M, Jeff Y
Team Leader: Tina E   Green Belt: Donna D
Team Members:
1. Mike S (ip)   8. (ED tech)
2. Stanalee G (patient access)   9. Materal management (PRN)
3. Glenda C (imaging)   10. Evening nurse
4. Bill N (MCE)   11.
5. David B (environmental)   12.
6. Vara S (MedSurge)   13.
7. Lori R (ED nurse)   14.

Event Date(s): May, 30 and 31, June 1, 2012   **Event Time:** Start: 0900   **End:** 1700

Event Location: Community room

Figure 8.6 Healing University's emergency department charter.

to review the respective responsibilities that your key leaders have assumed and determine, as a leadership team, whether any of the roles and responsibilities initially assigned should be revisited.

At a minimum, you will want to revisit the members and activities of the following teams:

▲ Transformation support team
▲ Healing pathway transformation teams
▲ Communication team
▲ Policy and procedure team
▲ Technical team
▲ Education team

You will also want to validate and get commitment to the schedule of meetings you have designed for these teams.

Step 9: Develop a Communication Plan

You have identified a communication team that will help develop and deliver messaging concerning your transformation efforts. But at this transformation summit you will want to openly discuss the issue to give team members guidance on how to proceed. Here are some key points to consider.

Most importantly, to what degree is the organization committed to diligently and rigorously engaging in, and completing, the transformation cycle? You don't want this to be the latest and greatest flavor-of-the-month program that is here today and gone tomorrow. Your team members will be skeptical, since they have seen other initiatives come and go. On one hand, the participation of your entire leadership team in this summit and the fact that you have gotten this far suggest that you have a very motivated and committed leadership team. On the other hand, most change initiatives do start out with enthusiasm and commitment. But they don't always sustain themselves. They start with the best of intentions, yet wither on the vine as other priorities intervene, or obstacles arise and are not readily overcome.

Thus, you may wish to consider a more moderate and subdued approach. Of course, your people need to know what the organization is up to. But perhaps rather than waving the banner and announcing a grand new program, one that is going to totally transform the organization (leading to immediate skepticism on the part of those who have seen such efforts start and fail), you might simply wish to describe what you are doing, without trying to build great expectations. In other words, you will be reviewing the key processes you have identified and looking for improvement opportunities. It is not a great new program; but rather a simple look at how you are providing patient care and seeing if you can do it better.

Then, after a few successful process transformations, an interesting phenomenon will start to occur: Those areas that have not yet participated will hear the buzz and the excitement and will be clamoring for the opportunity to be part of the effort. Lean thinking teaches us about the "pull" concept in a process. The same applies here: you want your team members to "pull" Lean transformation into their parts of the organization rather than push it on them.

Other issues to consider include the frequency of reporting out on team efforts and successes. Will you want to develop a transformation newsletter? Do you already have a regular publication in which you could add a transformation column? You will want to use print and social media to communicate success and maintain the momentum.

You have a team of smart people in the room with you during the transformation summit. Engage them in this discussion, and let them provide the communications team with enough to get started. You will adapt as you move forward.

CHAPTER 9

Ensure Enterprise Transformation Results

Enterprise transformation results can be measured in two ways: deployment and outcome. *Deployment* refers to the depth and breadth of the enterprise transformation across the organization. It is a measure of the maturity of the initiative. *Outcome*, as its name suggests, refers to the organizational impact achieved through the deployment effort.

The senior leadership team should review both deployment and outcome on a regular, ongoing basis. The objectives of this ongoing leadership role are to:

▲ Demonstrate by their participation and leadership that transformation is a priority
▲ Monitor transformation efforts and sustain results
▲ Identify and remove barriers
▲ Continue to reenergize the transformation
▲ Reprioritize efforts and initiatives as appropriate
▲ Identify new opportunities
▲ Standardize Best Practices

What Do I Do?

▲ Develop an enterprise dashboard.
▲ Develop healing pathway dashboards.
▲ Develop a tollgate assessment tool.
▲ Conduct regular 2- to 4-hour senior leadership team transformation support meetings to review outcomes and deployment.

How Do I Do It?

Step 1: Develop an Enterprise Dashboard

Most organizations have key measures they review on a regular basis. These should be organized into a single dashboard to be used for regular, at-a-glance monitoring of the overall health of the organization. The dashboard will be reviewed by the senior leadership team on a regular and ongoing basis, and it should include

▲ Metrics that reflect the overall health of the enterprise
▲ Metrics that reflect the outcomes of key transformation initiatives

The list below defines some metrics that might be useful for assessing the overall health of the enterprise. These are summarized in Table 9.1.

▲ **Unadjusted mortality rate:** A primary focus of patients, insurers, and public health agencies.
▲ **Significant adverse events:** Key quality-of-care metric.
▲ **Percent readmission ≤ 30 days/Medicare:** Factor that has significant impact on quality of care, costs, and patient wellness.
▲ **Hospital-acquired conditions (occurrences):** Conditions that have significant impact on quality of care and costs. The occurrences of such conditions identify potential clinical and operational opportunities for improvement.
▲ **Adjusted discharges:** Key indicator of clinical utilization.
▲ **Medicare case mix index:** Key indicator for Medicare volume and clinical utilization.
▲ **Total ALOS (CMI adjusted):** Proxy for hospital resources used per admission. Average Length Of Stay (ALOS) significantly longer than median group values may indicate operational inefficiencies. Note that while CMI adjustment improves the interpretability of the measure, better still is a measure that compares the system's actual patient population by Diagnosis-related group (DRG) to national ALOS norms based on that patient population.
▲ **Paid FTEs (including agency):** Measure of the staffing level the hospital maintains to serve its patient population.
▲ **Work FTEs:** Measure of the staffing level the hospital maintains to serve its patient population.

Table 9.1 Outcome Measures

Outcome Measures	Favorable Values
Clinical Quality	
Unadjusted overall mortality rate	Lower
Significant adverse events	Lower
Percent readmission ≤ 30 days/Medicare	Lower
Hospital-acquired conditions (occurrences)	Lower
Clinical Utilization	
Adjusted discharges	Higher
Medicare case mix index	Higher
Total ALOS	Lower
Labor	
Paid FTEs (including agency)	Lower
Work FTEs	Lower
Paid FTEs per adjusted discharge	Lower
Salary and wages as percent of net patient revenue	Lower
Overtime as a percent of salaries	Lower
Nonlabor	
Supplies and purchased service costs per adjusted discharge	Lower
Supplies and purchased service costs per adjusted patient-day	Lower
Drug costs per adjusted patient-day	Lower
Human Resources	
Benefits expenses per adjusted discharge	Lower
Salary and wages per adjusted discharge	Lower
Premium pay expenses per adjusted discharge	Lower
Financial Strength	
Net patient revenue per adjusted discharge	Higher
Total costs per adjusted discharge	Lower
Operating margin percent	Higher
EBIDA	Higher
Gross A/R days	Lower
Net A/R days	Lower
Cash as percent of net patient revenue	Higher
Debt service coverage ratio	Higher
Debt-to-capitalization ratio	Lower

▲ **Paid FTEs per adjusted discharge:** Key indicator for labor efficiency and benchmarking studies.

▲ **Salary and wages as percent of net patient revenue:** Measure of appropriate staffing levels based on revenue from operations.

▲ **Overtime as a percent of salaries:** Key indicator of labor efficiency.

▲ **Supplies and purchased service costs per adjusted discharge:** Measure of the proportion of a hospital's operating costs attributable to supplies and purchased services, on a per unit (case) basis.

▲ **Supplies and purchased service costs per adjusted patient-day:** Measure of the proportion of a hospital's operating costs attributable to supplies and purchased services, on a per patient-day basis.

▲ **Drug costs per adjusted patient-day:** Measure of the proportion of a hospital's operating costs attributable to pharmacy expenses.

▲ **Benefits expenses per adjusted discharge:** Measure of the proportion of a hospital's operating costs attributable to employee benefits.

▲ **Salary and wages per adjusted discharge:** Measure of the proportion of a hospital's operating costs attributable to employee labor.

▲ **Premium pay expenses per adjusted discharge:** Measure of the proportion of a hospital's operating cost attributable to employee labor at a higher wage rate.

▲ **Net patient revenue per adjusted discharge:** Measure of the average patient revenue per unit (per case) in hospital.

▲ **Total costs per adjusted discharge:** Measure of the average patient costs per unit (per case) in hospital.

HINT

It will be helpful for the enterprise dashboard to show metrics over time. A hyperlink to charts that visualize performance is even better.

▲ **Operating margin percent:** Measure of a hospital's viability absent investment income and charitable contributions; indication of the hospital's internal cash-generating ability to meet debt obligations.

▲ **EBIDA:** Proxy for cash flow.

▲ **Gross A/R days:** Measure of the amount of gross revenue that needs to be converted to cash.

▲ **Net A/R days:** Number of days of net operating revenue due from patient billings after deductibles for doubtful accounts.

▲ **Cash as a percent of net patient revenue:** Discrepancy between expected net revenue and actual net revenue. It may indicate payer shifts or opportunities for improvement in revenue cycle operations.

▲ **Days of cash on hand:** Number of days the hospital could operate if no further revenue were received. It is a measure of liquidity designed to reflect the hospital's ability to pay operating expenses with liquid assets.

▲ **Debt service coverage ratio:** Measure of the ratio of funds available to the year's principal and interest payment obligations. It is a proxy for creditworthiness.

▲ **Debt-to-capitalization ratio:** Measure of the hospital's debt load.

Additional patient care measures should also be reported on a regular basis, including the following *core measures*:

▲ Heart attack care
▲ Heart failure care
▲ Pneumonia care
▲ Surgical care
▲ Outpatient surgical care
▲ Outpatient heart attack care

In addition to capturing these *health of enterprise* (HoE) metrics, the dashboard should contain metrics associated with key healing pathway transformations currently underway. For example, if the emergency department (ED) is currently under transformation, the dashboard might include measures of ED length of stay, left without being seen percentage, or other appropriate metrics as determined by the senior leadership team.

Step 2: Develop Healing Pathway Dashboards

During the transformation summit, healing pathways were prioritized for transformation. Executive sponsors were assigned, and preliminary outcome metrics were identified. The healing pathway outcome metrics will be further developed and refined during the healing pathway analyses described later in this book.

For now, it is important to note that these healing pathway analyses and transformations will yield three key deliverables:

CASE STUDY

Healing University Healthcare System's Enterprise Dashboard

Outcome Measures	Favorable Values
Clinical Quality	
Unadjusted overall mortality rate	Lower
Significant adverse events	Lower
Percent readmission ≤ 30 days/Medicare	Lower
Hospital-acquired conditions (occurrences)	Lower
Clinical Utilization	
Adjusted discharges	Higher
Medicare case mix index	Higher
Total ALOS	Lower
Revenue Cycle	
Gross AR days	Lower
Cash as a percent of net patient revenue	Higher
Labor	
Paid FTEs	Lower
Work FTEs	Lower
Paid FTEs per adjusted discharge	Lower
Salary and wages as percent of net patient revenue	Lower
Overtime as a percent of salaries	Lower
Nonlabor	
Supplies and purchased service costs per adjusted discharge	Lower
Supplies and purchased service costs per adjusted patient-day	Lower
Drug costs per adjusted patient-day	Lower
Supply costs as percent of net patient revenue	Lower
Human Resources	
Benefits expenses per adjusted discharge	Lower
Salary and wages per adjusted discharge	Lower
Premium pay expenses per adjusted discharge	Lower
Financial Strength	
Net patient revenue per adjusted discharge	Higher
Total costs per adjusted discharge	Lower
Operating margin percent	Higher

CASE STUDY

Healing University Healthcare System's Enterprise Dashboard

Jul	Aug	Sep	Oct	Nov	Dec	YTD Actual	YTD Budget
3.14%	2.27%	2.20%	2.35%	2.30%	2.45%	2.45%	n/a
1	0	0	1	0	1	7	0
15.91%	15.33%	14.60%	14.34%	14.10%	14.15%	14.74%	17.00%
2	4	6	4	4	3	23	0
3,132	2,858	3,210	3,033	3,045	2,876	40,337	43,887
1.318	1.418	1.378	1.407	1.473	1.433	1.836	1.966
2.468	2.300	2.221	2.060	2.216	2.241	2.251	2.110
59.9	59.5	58.6	54.3	56.0	56.4	57.4	59.4
2.75	2.89	2.45	2.98	2.81	2.90	2.79	2.88
3,108	2,946	2,967	2,928	2,892	2,747	2,931	2,940
2,637	2,697	2,698	2,548	2,485	2,310	2,562	2,732
0.45	0.46	0.42	0.43	0.43	0.43	0.436	0.40
38.7%	39.4%	40.1%	38.4%	39.1%	39.4%	39.2%	40.0%
0.8%	0.7%	0.8%	0.8%	0.8%	0.8%	0.8%	0.8%
1,550.38	1,762.93	1,734.16	1,604.93	1,623.41	1,892.28	1,694.68	1754.6
297.33	344.93	342.53	331.85	315.14	375.02	334.47	342.6
54.81	66.60	61.01	56.82	45.21	58.22	57.11	50.00
8.6%	9.9%	9.4%	9.8%	8.6%	9.9%	9.4%	10.0%
1,125.62	1,066.31	1,264.52	1,011.95	1,020.77	1,008.53	1,082.95	1,025.00
4,452.16	4,414.20	4,713.75	4,285.62	4,420.64	4,310.42	4,432.80	4,800.00
195.46	180.54	171.43	165.19	164.00	211.93	181.42	190.00
13,516.24	13,529.68	14,481.22	12,618.20	13,612.38	14,317.03	13,676.63	13,800.00
12,269.73	12,780.95	1,3072.17	11,773.01	12,201.31	12,391.41	12,414.76	12,500.00
3.0%	3.5%	5.0%	4.4%	3.7%	4.0%	3.9%	3.1%

▲ A road map for transforming the healing pathway from the current state to a desired future state

▲ Identification of the outcome metrics that will be impacted

▲ Identification of process measures that will indicate whether the transformation is occurring as planned

To illustrate, Table 9.2 identifies four pathways that have been identified for transformation. Each includes a single sample outcome metric and a number of process metrics that might be associated with achieving the outcome measure. For example, one of the outcome measures for the emergency department is the average length of stay. The average length of stay may be impacted by a number of process measures, including time to triage, hours on diversion, and time to doctor.

The executive sponsor for each healing pathway will develop a dashboard of all *outcome* metrics for that pathway. Outcome metrics and key *process* metrics will be reported to the senior leadership team on a regular basis to ensure enterprise transformation results. More detailed process metrics will be reviewed at a lower-level team meeting, including the executive sponsor and his or her team, focused exclusively on the healing pathway they are charged with transforming. That process is discussed in Chapter 14 on ensuring healing pathway transformation results. A sample healing pathway dashboard follows, this one from Healing University's surgical services team.

Table 9.2 Healing Pathway Metrics—Drilling Down a Level from the Enterprise Dashboard

Healing Pathway	Sample Outcome Metric	Sample Process Measures
Emergency department	Average length of stay	Time to triage Hours on diversion Time to doctor
Revenue cycle	Gross A/R days	Percent bill holds > 5 days Percent orders not present
Surgical services	Cases per day	Percent on-time starts OR turnover time
In-patient care	FTE per adjusted discharge	Average length of intake Environmental Services on-time percent Nursing ratio

HEALING PATHWAY DASHBOARD

The surgical services transformation team at Healing University developed the following dashboard to track and report the results of its healing pathway transformation. Team members documented baseline and future state objectives, and they employed a simple red-yellow-green coding scheme to highlight progress (shown here in shades of gray):

▲ Red (dark gray): Inadequate or insufficient progress
▲ Yellow (light gray): Improving but not at goal future state
▲ Green (medium gray): Future state performance achieved

Metrics	Baseline	Goal	May	Jun	Jul	Aug	Sep	Oct
OR turnover time (min)	35	25	36	34	32	24	23	22
First case on time (%)	73	80	65	64	63	77	74	72
Worked hours/ 100 min.	14.2	12.28	13.4	13.4	13.4	12.24	12.0	11.98
Suite utilization 7 a.m.–11 p.m. (%)	60	75	61	61	57	104	86	84

Step 3: Develop a Tollgate Assessment Tool

The tollgate self-assessment is a great tool for analyzing Lean Sigma deployment. To conduct a tollgate assessment, evaluate the deployment of your Lean enterprise efforts across each of the areas of deployment.

The model proposed here identifies 12 areas of implementation that are critical to a successful enterprise transformation:

▲ Strategic plan and vision
▲ Aligning the organization
▲ Engaged leadership
▲ Enabling infrastructure
▲ Focus on value streams
▲ Rapid improvement events

▲ Engagement in projects
▲ Other Lean practices
▲ Six Sigma practices
▲ Partnering with suppliers/customers
▲ Innovative processes
▲ Pursuit of perfection

Maturity of deployment in each of the areas can be measured on a scale from 0 to 3:

▲ Tollgate 0
  ▼ Traditional non-Lean environment
▲ Tollgate 1
  ▼ Beginning to adopt the Lean paradigm
  ▼ Beginning to stimulate a cultural transformation
  ▼ Localized or limited improvement visible at value stream level only

TOLLGATE ASSESSMENT

This tollgate assessment is adapted from one developed by the U.S. Navy to assess its Lean Sigma deployment efforts.

▲ Tollgate 2
  ▼ Institutionalizing the Lean paradigm on a broad scale
  ▼ Broadly impacting the enterprise's culture of empowerment and improvement
  ▼ Positively impacting enterprise performance metrics
▲ Tollgate 3
  ▼ Self-sufficient Lean enterprise
  ▼ Cultural and performance transformation
  ▼ Operating performance at or approaching world class

A radar chart is an excellent tool for displaying the results of a tollgate analysis. The following sidebar illustrates the results of Healing University's tollgate assessment over a 4-year period.

2012 TOLLGATE ASSESSMENT

A summary of Healing University's current implementation status assessment is shown in the radar chart below. The analysis reveals an organization still in the middle stages of Lean implementation. Relative strengths exist in the areas of strategic planning and vision, alignment, rapid improvement events, and projects. Opportunities exist for focused implementation and rollout of Six Sigma practices, improved partnering with our suppliers/customers (physicians!), continued innovation, and pursuit of perfection.

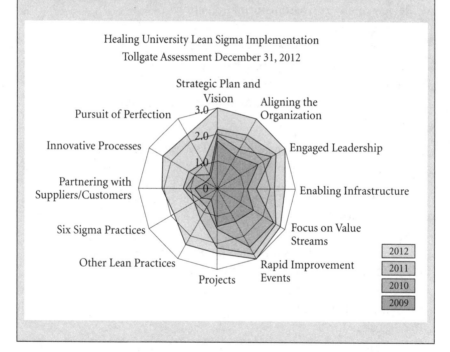

Healing University Lean Sigma Implementation
Tollgate Assessment December 31, 2012

Step 4: Conduct Regular 2- to 4-Hour Senior Leadership Team Transformation Support Meetings

This step is one of the most critical for ensuring the sustainability of the enterprise transformation effort. Continued participation of the senior leadership team in these regular and ongoing meetings will serve to maintain the drumbeat of the transformation.

It is recommended that these interactive, problem-solving sessions be held biweekly early in the transformation. They can be decreased to monthly sessions later, but delays in excess of a month may lead to failure of the transformation.

Each meeting should follow a standardized agenda, which might include the following items:

- ▲ CEO's welcome and remarks
- ▲ Review of actions from prior enterprise transformation meeting
- ▲ Review of enterprise operational dashboard
- ▲ Identification of opportunity areas and assignment of responsibility
- ▲ Review of enterprise transformation efforts
  - ▼ Deployment
  - ▼ High-level outcomes
- ▲ Review of each healing pathway transformation initiative
- ▲ Problem solving and elimination of barriers
- ▲ Evaluation and reprioritization of initiatives as appropriate
- ▲ Action item review

During the review of each healing pathway transformation initiative, the healing pathway executive should cover the following:

- ▲ Healing pathway dashboard
- ▲ Key accomplishments to date
- ▲ Barriers to success
- ▲ Any recommendations on the following elements:
  - ▼ Team structure
  - ▼ Team membership
  - ▼ Meeting frequency
  - ▼ Leadership
  - ▼ Work plan
  - ▼ Tracking and reporting needs
- ▲ Help needed from the senior leadership team
- ▲ Next steps

These transformation support meetings will maximize your organization's probability of success.

HINT

Your enterprise has a huge investment in the time and salaries of the senior leadership team. Maximize the effectiveness and efficiency of their time during these transformation meetings by providing an adequate level of administrative and analytic support:

▲ Project management and meeting support for the transformation
▲ Monitoring, tracking, and reporting on operational performance
▲ Monitoring, tracking, and reporting on pathway transformation efforts (help with dashboards)
▲ Analytic support
▲ Regular status reports on initiatives
▲ Maintenance of systemwide dashboard of all initiatives
▲ Comparative benchmarking activities and leveraging of best practices
▲ Action item capture, tracking, and follow-up

CHAPTER 10

Standardize Best Practices

"Standardize Best Practices" is the Act part of the Enterprise Transformation PDCA. As guided by this book, the senior leadership team (SLT) has:

▲ *Planned* the enterprise transformation at the transformation summit
▲ *Done* the transformation by transforming the healing pathways and value streams
▲ *Checked* the transformation through ongoing executive review sessions

Now the SLT will *Act* to eliminate obstacles and ensure the gains are sustained:

▲ Eliminate barriers.
▲ Ensure there is a stable quality management system (QMS) and a culture of compliance to enterprise policies and procedures.
▲ Ensure that new and improved processes are hardwired. This is only possible on the stable foundation of a culture of compliance.

What Is a Stable System?

A stable system is one in which there is consistent adherence to well-defined policies and procedures. The quality management system is typically the system designed to provide this stability. There are two fundamental pieces to the puzzle:

▲ Policies and procedures exist.
▲ They are followed.

The QMS is what provides the stable foundation. Many will be familiar with ISO 9000, the international quality management system (QMS) standard. Its basic requirement is to "Say what you do and do what you say."

> ## HINT
>
> A rigorous quality management system provides the stable foundation upon which a Lean transformation can thrive.

Any quality management system will identify similar requirements for the enterprise:

▲ Policies and procedures are well documented.
▲ People are trained to follow those policies and procedures.
▲ People have the necessary training and experience to do their jobs.
▲ An internal audit function exists to ensure consistent adherence to the documented policies and procedures.
▲ A corrective and preventive action system exists to track instances of noncompliance and ensure they are resolved.

> ## HINT
>
> ### Improving Patient Safety
>
> What is the cause of sentinel events? Often, it is a lack of adherence to a documented process. Sometimes the process is not documented; sometimes there is a human error; sometimes a willful neglect of following the agreed process. Institution of a rigorous quality management system will dramatically improve patient safety, regardless the cause.

Even in the absence of a Lean transformation, or any improvement effort for that matter, a rigorous quality management system will significantly improve patient safety and care while dramatically improving throughput and reducing cost.

System Out of Control

In the absence of a QMS, systems and processes will be out of statistical control. Inconsistent adherence to policies and procedures creates process variation that reverberates through the system, creating delays, dissatisfaction, and inefficiency.

As an example, consider operating room (OR) turnover. Let's say you collect 17 days of data on room turnover time in one of your surgical suites. You plot these and calculate the mean and control limits for the process, as shown in Figure 10.1. Each dot represents the turnover time after the first case for a given day. The average for those 17 days is 93 minutes, but there is significant variation from day to day: some much longer than 93 minutes and others less. Also a number of points are outside the control limits. This system is out of control, most likely due to a lack of standardized processes that are consistently followed.

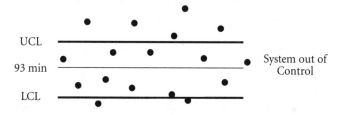

Figure 10.1   System out of control.

System In Control

Now imagine that you have standardized and consistently implemented the OR turnover process. You collect more data, plot them, and find that you now have a stable system, as depicted in Figure 10.2. To standardize processes is the way to achieve a stable system. Note that now there are no times outside the control limits. This means that the process is now more stable, more predictable. But you also get a bonus for your efforts. Notice that the average turnover time has also decreased somewhat, to 87 minutes.

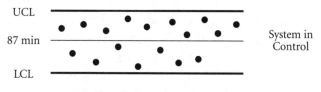

Figure 10.2   System in control.

Stabilizing a system usually results in some degree of improvement, even in the absence of intentional improvement efforts.

Improved System in Control

Once an enterprise has stabilized its processes, it can then successfully start to apply improvement methodologies. The aim is to achieve Best Practices consistently and predictably—to *Standardize Best Practices*. Then the control chart data for the sample OR turnover process might look something like Figure 10.3.

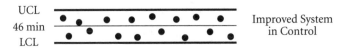

Figure 10.3   Improved system in control.

There are two things to note here. First, the average has been significantly decreased. Second, the control limits are much narrower than they were before. The process is improved and it is more predictable. This is the key to process, healing pathway, and enterprise transformation!

Successful Improvement Requires a Stable System

Deming said that before you try to improve a system, it needs to be stable. He said that trying to improve an unstable system was just "meddling." Trying to improve an unstable system is like building on sand. We need to treat our hospital system as we do our critical patients: We try to stabilize them first; then we look at improving their health.

Figure 10.4 illustrates the relationship between a stable system and process improvement. Let the PDCA wheel represent continual improvement. Continual turns of the PDCA wheel lead up the ramp toward ever-improving patient care and financial performance. The more continual improvement work, i.e., the more turns of the wheel, the better the performance. But standard policies and procedures, and the standardized processes they foster, are the chock block that helps hold the gains. Without

> ## HINT
>
> We should treat our hospitals as we do our patients: first stabilize; then improve health.
>
> For the healing enterprise that means to first stabilize with a quality management system and then improve with a Lean transformation.

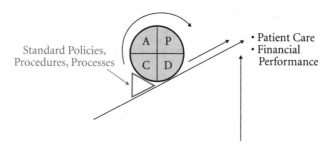

Standard Policies,
Procedures, Processes

A | P
C | D

• Patient Care
• Financial
  Performance

Figure 10.4   Relationship between a stable system and process improvement.

the organizational discipline of such standardized processes, an enterprise will not be able to hold the gains and the PDCA wheel will slip backward down the ramp.

In a nutshell, a stable system is really about eliminating, or at least minimizing, variation in a process. That is achieved through implementation of a sound quality management system. Then this stable system is improved by applying Lean Sigma to achieve continual improvement and ever-improving patient care and financial performance. This one-two punch will help any organization Standardize Best Practices to achieve perfect care.

How Do I Do It?

Step 1: Eliminate Barriers

The senior leadership team must eliminate barriers to the enterprise transformation. Many of these barriers will be identified by the healing pathway executives during the regular transformation support meetings discussed earlier in this book. These will be clearly identified barriers that the team can address head-on. Others will not be so obvious. The SLT will want to ensure an environment of open and honest inquiry to drive out fear and ensure that people are not afraid to bring problems forward.

Step 2: Ensure There Is a Stable Quality Management System

There should be an enterprisewide system and supporting infrastructure charged with the responsibility for ensuring a rigorous quality management system that fosters a culture of compliance.

A recommended infrastructure would include:

▲ A corporate office headed by an executive reporting to the CEO.
▲ Sufficient staff to support QMS efforts.
▲ A Quality Management Council comprised of senior leaders from operations and support departments.

The quality management system should include:

▲ Documented policies and procedures, preferably in an interactive online format accessible to all.
▲ A structure for reviewing and updating all policies and procedures.
▲ An internal audit system to ensure a culture of compliance with documented policies and procedures.
▲ A rigorous training and development program to address implementation of the documented policies and procedures.
▲ A corrective and preventive action system to capture actual and potential noncompliance and ensure appropriate action to preclude such occurrences.
▲ A continual improvement component to continue to identify and capitalize on new improvement opportunities.

Step 3: Ensure That New and Improved Processes Are Hardwired

Success with steps 1 and 2 above should ease the flow and maintenance of new and improved processes, policies, and procedures in the system. The senior leadership team should provide ongoing review and oversight to ensure the following:

- The documented quality management system is continually updated to reflect current practices.
- The QMS is responsive to both external requirements and internal improvements.
- Work instructions continue to be developed, documented, and updated at the process level.
- Policies and procedures continue to be developed, documented, and updated at the healing pathway level.
- New opportunities are identified and implemented.
- A culture of continual improvement permeates the enterprise.

CASE STUDY

Standardize Best Practices!

As part of their enterprise transformation efforts, the Healing University Healthcare System adopted the charge to Standardize Best Practices. Their CEO, Will Patterson, was the most vocal proponent and cheerleader for the transformation. His message to the team was clear and succinct:

> "We have to define the way we are going to do things, and ensure consistent adherence across the system. Once we have done that, we'll apply Lean Sigma to achieve best practices. That is the way we are going to reinvent Healing University."
>
> WILL PATTERSON, CEO
> HEALING UNIVERSITY HEALTHCARE

As Patterson explained, "We don't know what the Congress or the administration or the insurance companies are going to do. There is nothing we can do about these elements that are outside of our control. What we do know is this: the more we can improve the quality of care and do so at a lower cost, the better we will be. That is a constant. Standardizing Best Practices across Healing University is the single most important thing we can do to achieve those objectives. And Lean is the toolkit we will use to help us do that."

CASE STUDY (continued)

Over the course of Healing University's transformation initiative, they worked to standardize processes, policies, and procedures. Eventually, they began to call it "The Healing Way," i.e., the way we do things at Healing University. The double entendre was not lost on the team, as these new and improved processes also included the patient's healing experience.

The Healing Way: Standardize Best Practices!

PART III

Healing Pathway Transformation

CHAPTER 11

Plan the Healing Pathway Analysis

The healing pathway analysis (HPA) is the foundational element of process and cultural transformation for your healing enterprise. Through it, you will engage and empower a cross-functional team to evaluate and redesign your current processes. This will yield an improved value stream, providing better patient care at a lower cost. But it will also stimulate a cultural transformation to an empowered and highly motivated workforce.

At this point you should have prioritized the healing pathways and value streams you are going to transform. These may have been identified by your senior leadership team at a transformation summit, or you may have simply chosen a critical value stream to start with. In either case, you are now ready to begin the real work of transformation.

What Do I Do?

▲ Develop the team charter with the executive sponsor.
▲ Schedule and prepare for the HPA event.

How Do I Do It?

Step 1: Develop the Team Charter

The most important part of planning the HPA is formalizing the team charter. This document sets the expectations for team members and empowers them with the organizational authority to enact change. The team charter is the key tool for determining the membership and scope of the HPA. Depending on how you have gotten to this point, you may already have a draft. If not, you will want to develop one. Either you can develop it on your own and then meet with the executive sponsor and team leader to review and validate; or you can meet with the executive sponsor and team leader and develop it together. In either case, it is critical that you have complete executive sponsor support.

Charters come in different formats, but whatever the format, they should include the following information:

▲ The name of the healing pathway or value stream
▲ Purpose of the event
▲ Business impact
▲ Flow (process) boundaries
▲ Constraints
▲ Data needed beforehand
▲ Dates, time, and location
▲ Executive champion
▲ Team leader
▲ Team members

These are the key items you will want to finalize with the executive sponsor and team leader.

The charter that Healing University Hospital used for its emergency department HPA is shown in Figure 11.1.

TIP

Do not underestimate the importance of clearly defining and articulating the constraints. This tells team members what authority they have and consequently what authority they do not have. For example, are there limitations on the amount of capital dollars available to support the project? Full-time equivalents (FTEs)? Or will any requests for resources go through the normal approval channels? There is nothing more demotivating to a team than to come up with a set of recommendations that are denied due to constraints team members were not aware of. Clearly defined constraints will allow the team to come up with recommendations that are more likely to be supported.

Healing University Emergency Department

EVENT TYPE	[X] HPA	[ ] RIE	[ ] Project	[ ] Just Do It

Healing Pathway: Healing University Emergency Department

Event Purpose: Analyze and understand the current state. Design a desired future state and document a rapid improvement plan to get us there.

Business Impact: The ED is a front door to our system. Improving ED flow will result in reduced LWBS and LOS. Increased patient and team member satisfaction and increased OP and IP volume and revenue.

Flow Boundaries: Start point: Patient walks in the door   End point: Admit or discharge

Constraints: Any capital or other requests must be approved through normal channels. What can we do now with what we have? FTE not to exceed productivity targets.

Data: Data to collect beforehand: LWBS, hours on diversion, LOS, volume, percent ip/op, patients transferred to downtown, cover margin

Team Members: Executive Sponsor: Michael C   Black Belt: Floyd M, Jeff Y
(Name and Department) Team Leader: Tina E   Green Belt: Donna D
Team Members:
1. Mike S (ip)   8. (ED tech)
2. Stanalee G (patient access)   9. Materal management (PRN)
3. Glenda C (imaging)   10. Evening nurse
4. Bill N (MCE)   11.
5. David B (environmental)   12.
6. Vara S (MedSurge)   13.
7. Lori R (ED nurse)   14.

Event Date(s): May, 30 and 31, June 1, 2012   **Event Time:** Start: 0900   **End:** 1700

Event Location: Community room

Figure 11.1 Healing University emergency department charter.

The responsibilities of the executive sponsor and team leader were discussed briefly in Chapter 6 on governance, but they are elaborated on here, and you may wish to make these responsibilities clear during your meeting with them:

Executive Sponsor
▲ Has the authority to approve/disapprove process redesign changes
▲ Is accountable for team progress and removal of barriers

Team Leader
▲ Ensures that the objectives of the team are met
▲ Ensures that team tasks are appropriately assigned to team members
▲ Establishes the team meeting content, agenda, date, time, and location
▲ Follows up on members' efforts and commitments
▲ Manages conflicts to resolution
▲ Participates as a team member
▲ Summarizes and confirms commitments, decisions, and outcomes of the meeting

Team Members

▲ Learn and apply team and problem-solving skills

▲ Actively participate

▲ Take on and carry out assignments

▲ Communicate the team's activities to other members

▲ Communicate other members' ideas to the team

▲ Support team decisions in word and deed

▲ Adhere to team ground rules

Step 2: Schedule and Prepare for the HPA Event

Now that you have executive sponsor support, you are ready to schedule and prepare for the HPA. The typical length of an HPA is three full days. Coordinate with your executive sponsor and get invitations out to all participants as far in advance as possible.

The following items will be needed to help facilitate a smooth event.

1. A minimum of two flip charts is needed in the room.
2. An assortment of markers is needed for the flip charts and cards.
3. A laptop and projector are needed.

IMPORTANT

You will want to have multiple copies of the floor layout of the area(s) for which you are conducting the HPA. Your plant engineering team will be able to provide these to you.

HINT

With regard to location, you will want something on-site or very close to the area for which you will be conducting the HPA. The reason is that you will spend a lot of time on the floor, actually walking and examining the process.

HINT

The ideal space for conducting the event will allow your participants to sit either at a large executive table or in a U shape to enable interaction and dialogue. You will want a lot of wall space since most of the work, including process mapping and team problem solving, is done "at the wall." A typical room layout is shown in Figure 11.2.

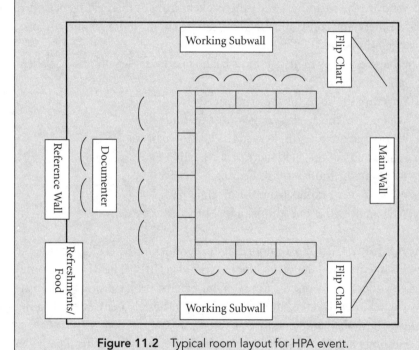

Figure 11.2   Typical room layout for HPA event.

4. Assorted sticky notes and cards will be helpful.
5. It is wise to have lunch provided in the room along with morning and afternoon beverages. This will keep the group working throughout the day.
6. A camera is needed to capture the team's work so that it can be documented.
7. Multiple copies of the layout of the area are useful.
8. If possible, prework can be started. See the prework list.

In preparing for this event, you may wish to give consideration to the following questions. Many of these will be developed by the team during the event, but advance consideration by the facilitator and team leader may provide helpful hints to the team:

1. Why was this healing pathway chosen (patient issues, financial considerations, lead time, labor content, etc.)?
2. What product/service does this healing pathway provide?
3. Who are the customers and other stakeholders of this healing pathway?
4. Are there any constraints to what can be changed due to current laws, regulations, etc.?
5. What are the key metrics for this healing pathway (length of stay, patient satisfaction, etc.)?
6. What are the goals for these metrics?
7. Where does this healing pathway begin and end?
8. Who/what are the suppliers to the healing pathway?
9. Who/what are the customers of the healing pathway?
10. What are the inputs to this healing pathway?
11. What are the outputs from this healing pathway?
12. What are the five to eight principal process steps?

The Agenda. The most important deliverables of the HPA event are the future state map and the *rapid improvement plan* (RIP). The RIP is the time-phased collection of rapid improvement events, projects, and Just Do Its necessary to develop the future state. Every other element of the healing pathway analysis is aimed at those two ends.

Healing pathway analyses can be very fluid events. The critical path will usually be the same, but the information-gathering and problem-solving tools employed may vary depending on the circumstances. Here are the key elements that will be common to nearly every healing pathway analysis:

▲ Suppliers-inputs-processes-outputs-customers (SIPOC) diagram
▲ Current state
▲ Ideal state
▲ Future state
▲ Rapid improvement plan

Depending on how savvy the team is about quality improvement, you may do some teaching on Lean, team problem solving, or other Quality Improvement tools. The tools you employ will vary. The time it takes to complete the different phases of the HPA will also vary. The key is to follow the critical path, but acknowledge that there is no clear time line to which an HPA team will adhere. You will depend on the skills of the facilitator to guide the process as needed.

Here is a moderately detailed agenda you may wish to use as a starting point. Adapt it to meet your needs, and recognize that it is only a plan!

Day 1
▲ Morning
  ▼ Make introductions.
  ▼ Begin healing pathway analysis training.
  ▼ Review types of process maps.
  ▼ Develop the SIPOC diagram.
  ▼ Identify key metrics.
  ▼ Define boundaries.
  ▼ Define the value.
▲ Afternoon
  ▼ Set up the process map.
  ▼ Conduct process walk.
  ▼ Gather data.
  ▼ Draft current state map.
  ▼ Develop visual maps (spaghetti map, circle diagram).

HINT

An experienced facilitator is worth his or her weight in gold when it comes to ensuring a successful healing pathway analysis. A Lean expert with healthcare experience will maximize your probability of success.

Day 2
▲ Morning
  ▼ Review the previous day's activities.
  ▼ Review waste walk and healing impediments walk techniques.

▼ Organize into several teams as necessary with "shotgun" launch of waste walk.
▼ Debrief the waste walk.
▲ Afternoon
▼ Organize into several teams as necessary with shotgun launch of healing impediments walk.
▼ Debrief the healing impediments walk.
▼ Develop the ideal state.
▼ Identify guiding principles.

Day 3
▲ Morning
▼ Review the previous day's activities.
▼ Develop the future state.
▲ Afternoon
▼ Identify and categorize opportunities for improvement.
▼ Develop a rapid improvement plan.
▼ Conduct an out-brief to the executive sponsor.
▼ Celebrate success.

With this planning complete, you should be ready to conduct your healing pathway analysis!

CHAPTER 12

Healing Pathway Analysis Part I: Understand the Current State

This chapter describes the steps involved in fully understanding the current state of the healing pathway. The setting is the time and place of your healing pathway analysis. It has taken a lot of legwork and planning to get to this point. If it is well facilitated, the team will find this a very engaging and rewarding experience.

What Do I Do?

▲ Introduce the event and the participants.
▲ Conduct appropriate training.
▲ Develop the SIPOC diagram.
▲ Define *value* and *expected outcome*.
▲ Set up the map.
▲ Walk the process and collect data.
▲ Develop the current state value stream map.
▲ Develop visual maps.
▲ Conduct a waste walk.
▲ Conduct a healing impediments walk.

How Do I Do It?

Step 1: Introduce the Event and the Participants

The executive sponsor should welcome the team to the event. She or he might consider:

- ▲ Thanking the members for their participation
- ▲ Highlighting the importance of the healing pathway they are working on and the benefits of improving it
- ▲ Challenging them to be very creative in their thinking and to design a process that will improve patient care and operational performance
- ▲ Pledging his or her support to the team's efforts, and promising to stay engaged and help remove barriers to the team's progress

Next, the team leader should review the team charter. This will get everyone on the same page regarding the team's objectives and the desired outcomes, thus ensuring that all participants are at the same starting point when they start to develop the SIPOC diagram.

Team introductions could be conducted in any number of ways—choose your favorite icebreaker or opener. Sometimes it is effective to use a soft rubber ball or other object that can be passed around. The facilitator makes the first introduction and then passes the ball to someone who goes next. When that person is done, she or he throws the ball to someone else until everyone has a turn.

Step 2: Conduct the Appropriate Training

Whether this is your first or fortieth HPA, it is likely that some of the participants will not have prior experience. For that reason, it is helpful to provide some high-level training. In addition, this information will serve as a refresher to those who have participated before. Keep in mind that HPAs are very informal and very interactive, so you will do a lot of teaching "just in time" along the way. However, as an introduction, you might wish to consider the following:

Healing Pathway Analysis Training

▲ Value-added versus non-value-added definitions
▲ The eight wastes
▲ The eight impediments to healing
▲ "Flow" and "pull" systems
▲ Creating flow: batch and queue versus one-piece flow
▲ 5S (sort, set-in-order, shine, standardize, sustain) overview and visual management
▲ SIPOC diagrams
▲ Flowcharting
▲ Waste map
▲ Deployment flowcharting
▲ Value stream mapping
▲ Definitions of *cycle time, first-time quality, just in time* (JIT), and *mistake proofing*

Team Roles

▲ Executive sponsor
▲ Facilitator
▲ Timekeeper
▲ Recorder/scribe
▲ Members

Finally, the team should review and commit to the team ground rules. These ground rules are intended to facilitate open and honest discussion in an environment of respect and inquiry. Your organization may have a set of ground rules that it likes. If so, use it. If not, consider using that used by Healing University Healthcare System, shown in Table 12.1.

Step 3: Develop the SIPOC Diagram

SIPOC is the acronym for suppliers-inputs-process-outputs-customers. The purpose of developing a SIPOC diagram is to gain a high-level understanding of the process, identifying all relevant elements under consideration for the improvement effort. An example is provided in Figure 12.1.

Table 12.1 Healing University Team Ground Rules

Healing University Healthcare System Ground Rules for Team Events
▲ Keep an open mind to change.
▲ Maintain a positive attitude.
▲ Decision making is done by consensus.
▼ A decision everyone can live with
▼ Not a vote
▼ Never leave in silent disagreement
▼ One person, one opinion, no position or rank
▲ Start and end on time.
▲ There should be only one conversation at a time.
▲ All parties should show mutual respect.
▲ Build on ideas — do not tear them down.
▲ Leave baggage outside the door.
▲ Create a blameless environment.
▲ Participation requires
▼ Attendance
▼ Completion of assigned work
▲ There is no such thing as a dumb question.
▲ Las Vegas rules: What happens in the room stays in the room. It can be disastrous if misinformation is leaked from the room.
▲ Parking lot: The parking lot is a means to keep everyone focused but not to dismiss any ideas. If someone brings up something that does not pertain to the value stream, it is captured on the parking lot flipchart for review at a later date.

Step 4: Define Value and Expected Outcome

To better understand the purpose of the healing pathway—or any value stream for that matter—the value to the customer needs to be determined. In the healing pathway examined, what is it that the patient wants from the total healing pathway? (For example, in the emergency department, value to the customer may be quick and expert treatment.) The team should address the following questions:

▲ Who is the primary customer?

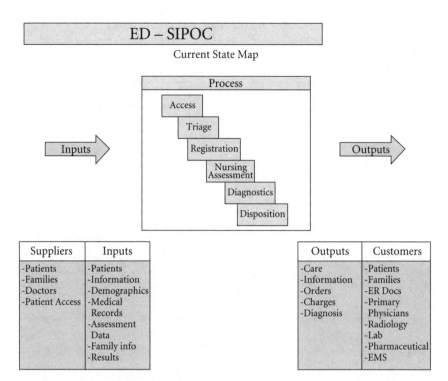

Figure 12.1   Healing University emergency department SIPOC diagram.

▲   What is the value added?
▲   What value is the customer willing to pay for?

The business case should also be articulated and documented by the team. An HPA is a huge investment of time and energy by the hospital and by every member of the team, and only high-yield opportunities merit such attention. How will this benefit the organization?

Lastly, what are the key outcome measures that the team will use to measure success? The key metrics identify the means to measure improvements within the healing pathway. These are what the HPA will review going forward, and these measures will be rolled up to the senior leadership team for ongoing review.

Healing University Healthcare System's business case and key outcome measures for the emergency department healing pathway analysis are shown in Table 12.2.

Table 12.2 Healing University Healthcare System's
Business Case and Key Outcome Measures

	Goal	Result
Business Case	Define the business case for the HPA. Why are we here? How will this benefit us?	Improve patient satisfaction Improve quality/effectiveness of care Reduce stress on patients and staff Increase/maintain volume Maximize use of space Physicians allowed more contact time with patients Reduce staff turnover Increase staff satisfaction
Key Requirements (musts)	Requirements that can't be changed	Triage is a function not a place Level I and II patients ID's and pull through Bedside registration Dedicated x-ray resources—people and equipment Lab and x-ray for ER treated as priority Improve flow for patients and care keepers Adequate resources in the ER (supplies/meds/etc.)
Key Measurements	Two to three metrics which will measure future success	Total LOS Door to decision Boarding time after decision Order to receipt Order to result Door to room Door to physician Patient satisfaction

Step 5: Set Up the Map

The purpose of this step is to add sufficient clarity to the process such that the team can conduct its initial process exploration. You will begin with a simple process map. The high-level process you have identified in the SIPOC diagram is a good starting point. Drill down another layer below that, as illustrated in Figure 12.2.

Within clearly defined boundaries

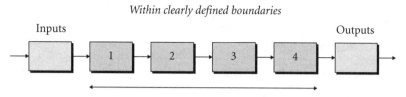

Figure 12.2   Identify the process steps.

Step 6: Walk the Process and Collect Data

Observe the value stream in action. Talk to the nurses, technicians, physicians, and other team members as you walk through the process. Ask them what types of problems they see on a regular basis. They will be a rich source of process improvement knowledge. Interview team members from different shifts and similar service operations, as appropriate. Verify your observations against documented policies, procedures, work flows, and job aids. During data collection, capture the cycle time of each process step. You will use this information to create a lead-time chart at the bottom of your value stream map. Determine lead times for value-added and non-valued added process steps. Determine what the trigger to the process step is and how one knows it is complete.

POINT TO PONDER

Any process has at least three versions:

What you think it is
What it actually is
What it should be!

Don't rely on memory alone. Give each member of the team a clipboard and a stack of process sheets. A sample is provided in Figure 12.3.

Identify activities that have queues. Identify all activities where the patients, associates, or work must wait in a queue. Indicate how the queues of work are prioritized and sequenced.

If the activity requires the use of software applications, perform sequential screen captures as a way to document the flow of tasks and information.

Step Number:	Step number of this step in the process
Process Description:	
Description of process being performed:	
Demand	How many items the customer wants per ___
Trigger	What tells you to begin this step?
Done	What tells you the step is finished?
Flow Time	Entire time it takes to complete the step?
Touch Time	Actual hands-on time required to complete the step
People	Number of people required to complete the step
Stations	How many places this step is performed
WIP	How many items have been started but not completed
WIQ	How many items are waiting to start?
FPY	Percentage of defect-free items delivered to the customer the FIRST TIME
Changeover/Setup	Time between the last step of the current job to first step of next job
Flow Stoppers	Problems that keep you from doing your job

Figure 12.3 Sample process data sheet.

Record exactly what you see without making any judgments. There is no right or wrong—just record what is actually happening.

A sample ED data collection sheet is shown in Figure 12.4.

> ## TIP
>
> Make note of whether activities are performed sequentially in real time or in batches. Record if the activity requires multiple resources to review or approve before proceeding to the next activity. If activities are performed in batches, show the size of the batches and how often they occur. If the activity must go through multiple resources for review or approval, indicate average process delays.

The team was interested in determining the average time patients spent in the ED on a given day. They collected data on a number of patients, using the following data collection sheet:

Diagnosis
Total process time: 59 minutes
Time patient waited: 29 minutes (49 percent)

Step	Time In	Time Out
Sign in	1527	1527
Triage	1527	1530
Registration	1533	1544
Go to room	1530	1623
Nurse assessment	1533	1538
MD/PA assessment	1540	1548
Urine	—	
Lab	1540	1555
CT	—	
Treatment	0	Rx
MD/PA decision	1600	1605
Feedback to patient	1610	1615
Disposition	1620	1623
Room turn around	1624	1626

Over the course of the day, the team determined that the average cycle time was 1 hour, 38 minutes, of which 45 minutes, or 46%, was time waiting.

Figure 12.4   Sample ED data collection sheet from Healing University.

Step 7: Develop the Current State Value Stream Map

A value stream map uses simple graphics or standardized icons to show the sequence of activities, the flow of information, the identification and alignment of resources, the identification of value creation and value destruction (waste), and the lead time of specific value stream activities. *The Lean Memory Jogger for Healthcare* is a great resource for value stream mapping in healthcare.

Complete the map in two phases. First, use adhesive notes that can be easily rearranged while your team comes to a consensus, or use a pencil and eraser to draw and refine your map. This is the version on which the team will place the appropriate dots to denote value-adding, non-value-adding, and non-value-adding but required steps.

An example of what such a map may look like when documented is provided in Figure 12.5.

Each process in the map can be identified as value-added, non-value-added or non-value-added but necessary. The participants are to mark the data boxes with a green, red, or yellow dot. The green dots represent the value-added steps. The red dots represent pure waste, and yellow dots

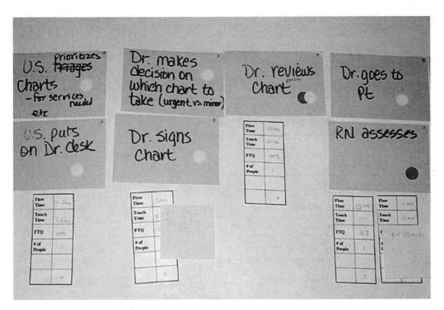

Figure 12.5 Example of current state value stream map.

HINT

The Lean Memory Jogger for Healthcare contains more detailed information for developing a value stream map including a summary of all mapping symbols.

---

represent the non-value-added but required steps. The majority of the steps will likely be red and/or yellow. After each has been marked, the team or a member from the team should add up the total value-added time and the total non-value-added time.

Expect individual differences in perception of reality. These differences are not bad at the beginning of the mapping process, but should not exist at the end. If they do, then they need to be resolved as the team compiles and integrates various views of how activities are accomplished. The objective is to develop the most accurate map of the value stream possible.

Next, finalize the value stream map. A sample map is provided in Figures 12.6 and 12.7.

Last, calculate the overall performance of your healing pathway. This will be an effective baseline measurement against which to compare future improvements. See Table 12.3 for the summary statistics from Healing University's current state analysis.

Step 8: Develop Visual Maps

The term *visual map* refers to illustrations that enable the team to clearly visualize the flow of people or information in the process. They may include spaghetti diagrams, circle diagrams, or inventory control maps.

Table 12.3   Healing University ED Current State Analysis

Performance Data	Current State
Total cycle time	258.9 minutes
Total value-added time	55.8 minutes
Total non-value-added time	203.1 minutes
Value added	21.6%

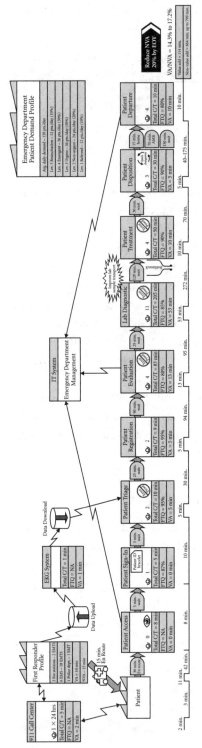

Figure 12.6 Emergency department value stream map.

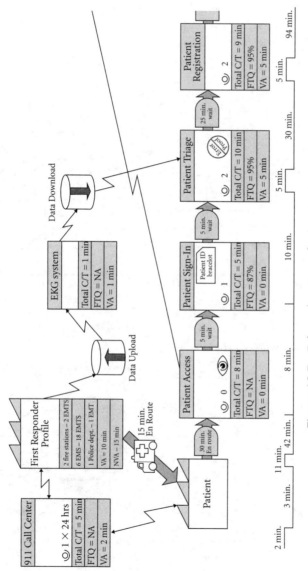

Figure 12.7 Enlarged portion of value stream map.

117

A spaghetti diagram is most commonly used, and it is a very useful graphic that shows how people or things flow in a layout. On the layout, lines show the flow of the patients, staff, equipment, or materials through the facility. It is meant to identify excess transportation or motion. Use the copies of the floor layouts you received in preparation of the event to map the flow of patients, staff, and equipment as they apply to your analysis. If these are not available, draw your own. See Figure 12.8 for a sample.

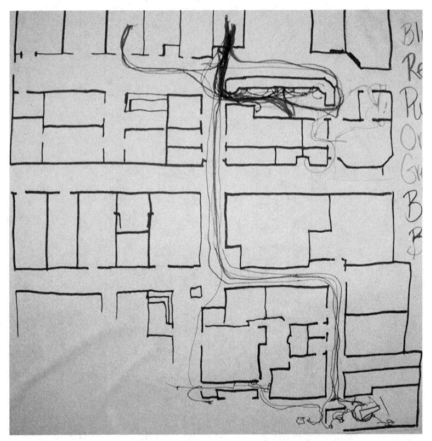

The lines in the spaghetti diagram represent the paths traveled by the people involved in the Emergency Department Current State Process. Each color represents a different contributor in the process:

Contributor	Color	Contributor	Color
Registraton Clerk	Green	Doctor	Brown
Triage Nurse	Orange	X-ray Technician	Purple
Nurse	Blue	Unit Secretary	Black
Ed Technician	Red		

Figure 12.8 Healing University's ED spaghetti diagram.

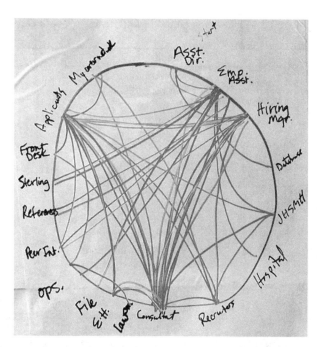

Figure 12.9   Healing University's circle diagram for ED analysis.

A circle diagram is used to identify unnecessary processing. On a large sheet of paper a circle is drawn, and the name or title of every person who touches the process is written on the outside of the circle. A line is drawn on the diagram to show how the process flows from one person to the next.

Figure 12.9 shows Healing University's circle diagram for ED analysis.

Step 9: Conduct a Waste Walk

During this walk through the process, the team will be looking for evidence of any of the eight wastes. Remind the team of what they are; or better yet, provide them with a handout to carry on their walk. Table 12.4 provides the information for such a handout. If the team is relatively large, then a shotgun start (in which the team breaks into smaller groups and starts at different stages of the process) may be appropriate.

Team members should observe and learn about the process and write down observations and opportunities, one each, on index cards. These will be further analyzed during the event. Upon reconvening, all participants

Table 12.4 Examples of the Eight Wastes in Healthcare

Waste	Result
Defects	Medication error
	Wrong patient
	Wrong procedure
	Missing information
	Redraws
	Poor clinical outcomes
Overproduction	Drawing too many samples
	Extra tests
Waiting	Waiting to see a doctor
	Waiting for a procedure
	Waiting for a bed
	Waiting for lab results
	Waiting for discharge
Not utilizing people's abilities	No empowerment
	"Check your brains at the door" mentality
	Old guard thinking, politics, or the culture
	Poor hiring practices
	Low or no investment in training
	High turn-over
Transportation	Moving samples
	Moving specimens
	Moving patients for testing
	Moving patients for treatment
	Moving equipment
Inventory	Pharmacy stock
	Laboratory supplies
	Samples
	Specimens waiting analysis
	Paperwork in progress
Motion	Searching for patients
	Searching for medications
	Searching for charts
	Gathering tools
	Gathering supplies
	Handling paperwork
Excess processing	Multiple bed moves
	Multiple testing
	Excessive paperwork
	Unnecessary procedures

should discuss the opportunities (problems) that they observed, and their index cards will be captured on the wall for further analysis. The information collected during the waste walk is crucial in informing the actions needed to achieve the future state.

Wastes identified in the Healing University ED are shown in Table 12.5.

Table 12.5   Wastes Identified in the Healing University ED

Waste	Result
Defects	Same patient labeled with two persons' labels (in lab)
	Labels for lab not segmented for different patients—lab will not accept
	Blood specimens not labeled (nurses' station)
	Slow log-in; cumbersome for users
	ER access signage is bad/nonexistent both on and off campus
	Lack of parking for ED
Overproduction	Clipboards for charts redundant
Waiting	Delay for MD assessment on employee care side of ER
	Delay in report for admissions because of private MD orders
	Patients waiting over 30 minutes to see MD
	Delay in MD assessment while MD waits for chart to be placed in "MD to see" rack
	Patients holding assigned to clean beds
	Delay in lab collection
Not utilizing talent	Unused protocols
	Nurse communication
	Triage room leaving area to find nurse to give report
	Medical imaging techs transporting their own patients
	Most ER physicians not taking advantage of wireless capability
	Triage nurse leaving triage room to see if a room is empty and clean
	No consistent use of orders protocols
	Lab techs—area for lab draws backed up—rely on ER technicians to prioritize

(continued on next page)

Table 12.5 Wastes Identified in the Healing University ED (*continued*)

Waste	Result
Transportation	Time delay due to slow transport
Inventory	Delays with MD on employee-care side
	Better access to all supplies and medications in observation beds
Motion	Lab label printer and labels not conveniently located
	Triage nurse leaving triage room to see if room is empty and clean
	Multiple lab specimens in several different locations
	Laundry cart has to be moved to get stretcher in X-ray room
Extra processing	40% of room assignments are changed
	MDs have individual processes for their steps
	Multiple processes for identifying patients at access points
	Duplications and greeter's functions
	Patient taken to triage when bed available
Other observations	Congested space at the nurses' desk
	Cluttered hallways
	A lot of "that's not my patient"
	Triage nurse not compliant with uniform policy—shirt and gum
	Blood draw label and depository disorganized
	Blood collection prior to MD orders being placed
	Blood tubes in bag by room
	Inconsistent job specifications and training—all disciplines
	No "fast track" for nonemergency patients
	Too few parking spaces for ER patients
	Easy to get lost trying to find ER
	Better signs for ER

Step 10: Conduct a Healing Impediments Walk

In the case of a healing pathway analysis, the team will also want to conduct a healing impediments walk. During this process exploration step, the team will be looking for evidence of any of the eight impediments to healing. Remind the team of what they are; or better yet, provide them with a handout to carry on their walk. Table 12.6 could be such a handout. If the

Table 12.6   The Eight Impediments to Healing

Impediment	Description
Stress and anxiety	A common state of patients created by the malady that brought them to us, uncertainty and fear about what may be wrong with them, and unfamiliarity with our setting
Inactivity and waiting	Idle and unproductive time created when we cannot tend to our patients at a rate appropriate to their treatment
Coldness or apathy	An aloofness or distancing from the patients by one of his or her caregivers
Knowledge gap	The lack of information patients have about what is wrong with them, what is happening to them, and what is going to happen when
Neglect	The absence of steady interaction and information sharing with the patient
Embarrassment	A negative patient experience casued by a lack of dignity in the treatment process
Submission and helplessness	A state of learned helplessness exacerbated by information, power, and social status differentials
Statistic	Depersonalization of patients: the chest pain in room 13, 5 boarders in the ED

team is relatively large, then a shotgun start (in which the team breaks into smaller groups and the groups start at different stages of the process) may be appropriate.

As with the waste walk, team members should observe and learn about the process and write down observations and opportunities, one each, on index cards. These will be further analyzed during the event. It is helpful to have someone become a patient and proceed physically through the process, trying to observe it from the patient's perspective. However, be mindful that a known employee will likely be treated differently by the staff than a patient would. For that reason, it is critically important to solicit real patient input during the impediments walks. Obviously, you will want to use best judgment as to whether it is appropriate to engage with any given patient. But to the extent possible, the patients' direct input will be a rich source of data as you try to identify the eight impediments to healing. Upon reconvening, all participants should discuss the opportunities (problems) that they observed, and their index cards will be captured on the wall for further analysis. The information collected during this walk is also crucial in informing the actions needed to achieve the desired future state.

The impediments to healing identified during the Healing University ED healing pathway analysis are provided in Table 12.7.

Table 12.7 The Impediments to Healing Identified During the Healing University's ED HPA

Impediment	Description
Stress and anxiety	Patient looked scared There was lots of commotion in the ED Patient said she was afraid because she didn't know what was wrong with her
Inactivity and waiting	ED waiting room was backed up with several patients Three patients were waiting for a room upstairs
Coldness or apathy	Doctor seemed distant when talking with patient in room 4
Knowledge gap	Patients said they were supposed to have an MRI but didn't know when it would happen Patients didn't know if they were done and free to go or if they were still waiting for results
Neglect	Patient was in room 3 for over an hour with no update or progress
Embarrassment	Patient had to go to the bathroom but was uncomfortable doing so in the hospital gown Patient was uncomfortable with being in a bay "open to the world"
Submission and helplessness	Patient in room 12 indicated she was afraid to ask for additional information and clarification when she didn't understand what her physician told her
Statistic	None observed

CHAPTER 13

Healing Pathway Analysis Part II: Design the Future State

Grounded in a thorough understanding of the current state, the team is now ready to develop the future state for your healing pathway or value stream.

What Do I Do?

1. Envision an ideal state.
2. Capture the guiding principles of the ideal state.
3. Conduct group problem solving around waste and current impediments to flow and healing.
4. Design an achievable future state.
5. Identify and prioritize opportunities.
6. Develop a rapid improvement plan.

How Do I Do It?

Step 1: Envision an Ideal State

This is the "breakthrough" portion of the event. To get the group members thinking "outside the box," it is necessary to have them identify an ideal state. Depending on the group, the facilitator, and how well the event is going, you should break the group into smaller teams and give each team about an hour to create an ideal condition. They should develop both a layout and a map. Each group will present its work to the larger team to further stimulate ideas and creativity.

The team members should be relatively unconstrained during this exercise. You may simply instruct them as follows: Assuming that there were no constraints on expenses, capital, or people, and that you could magically wave a wand to redesign it any way you wanted to, how would you design it?

Figure 13.1 shows two versions of an ideal state emergency department as designed by the Healing University's ED team.

Step 2: Capture the Guiding Principles

After the teams create an ideal state and provide respective out-briefs, the entire group should brainstorm to identify the guiding principles they employed in their designs. The guiding principles are the items identified as key themes used in developing the ideal state. A typical list may include the following:

▲ No waiting
▲ One-piece flow
▲ Fully informed patient
▲ Flexible staffing
▲ Guaranteed patient privacy
▲ Good teamwork
▲ Physician participation
▲ Good communication
▲ Patient focus
▲ Efficient operation
▲ Standardization
▲ Ownership by all
▲ Idea that functions are not places
▲ Paperless operation
▲ Use of mobile and wireless technology
▲ "Pulling" the patient
▲ No waste
▲ Standardized processes

These guiding principles should be designed into the achievable future state, to the extent possible.

During Healing University's ED healing pathway analysis, the group broke into two teams to develop an ideal state. As can be seen in the drawings, the designs shared very similar characteristics:

▲   A circular suite of patient rooms

▲   Centralized nursing station for ready viewing and access to each patient room

▲   Family access from the opposite side of the patient's room

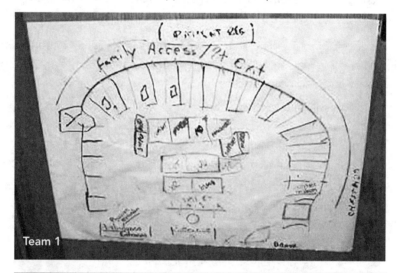

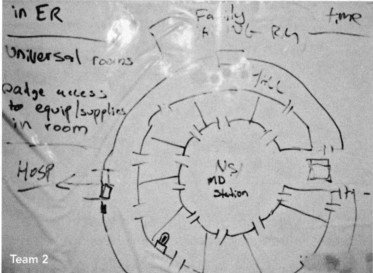

Figure 13.1   Ideal state emergency department designs.

The following case study shows the guiding principles developed by the ED team.

CASE STUDY

Guiding Principles for an Emergency Department at Healing University Hospital

After developing an ideal state model of the emergency department during Healing University's ED healing pathway analysis, they captured the following guiding principles:

- ▲ Clinical value added at front
- ▲ Paperless operation
- ▲ Planned, call-ahead service
- ▲ Minimize staff and patient walking
- ▲ No waste
- ▲ Standardized processes
- ▲ No waiting
- ▲ Floors begging for patients
- ▲ Fast, efficient, effective operation
- ▲ Happy staff
- ▲ Patient expectations consistently exceeded
- ▲ Idea that functions are not places
- ▲ Easy, fast access to outside services
- ▲ Patients flow
- ▲ Secure and safe operation
- ▲ Good communication
- ▲ Flexible, user-friendly systems
- ▲ Multiskilled employees

Step 3: Group Problem Solving Around Waste and Current Impediments to Flow and Healing

During the current state analysis, team members collected a wealth of data. This information has taken a number of forms, including impediments to healing from the patient's perspective, wastes in the process, impediments to process flow, and quantification of non-value-added and non-value-added but required process elements. The purpose of the current activity is to develop those data into actionable information.

Depending on the facilitator, the team, and the type of data being evaluated, there are a variety of ways to develop actionable objectives for improvement. For example, the opportunities may be organized using an affinity diagram and prioritized for resolution; they may be subject to the five whys; or they may benefit from a force field analysis. A Multivoting technique, such as the previously described 3-2-1 voting process, may also be an appropriate activity. Finding solutions to these challenges will yield great improvements to the healing pathway in terms of patient experience, throughput, quality, and cost.

The team may wish to give consideration to the Eight Enablers identified in Table 13.1. These will help counter any of the eight impediments discovered during the flow analysis.

Table 13.1   The Eight Impediments and the Eight Enablers

Impediments	Enablers
Stress and anxiety	Calm and comfort
Inactivity and waiting	Progress
Coldness or apathy	Caring and warmth
Knowledge gap	Abundant communication and understanding
Neglect	Engagement
Embarassment	Dignity
Submission and helplessness	Respect and empowerment
Statistic	Special

> ## HINT
>
> Regardless of the group problem-solving technique you choose to use, the following recommendations will be helpful:
>
> ▲ Everyone participates.
> ▲ Take turns.
> ▲ Be creative; do not hold back.
> ▲ Build on the ideas of others.
> ▲ Challenge assumptions.

> ## HINT
>
> Group brainstorming should occur in two phases. In phase I, the generation of ideas and "outside the box" thinking should be encouraged. During this phase, there should be no criticism of ideas. In phase II team members select only those ideas that they determine have the greatest probability of success. This phase requires objective evaluation of the strengths and weaknesses of each proposed solution.

The case study, "Improvement Opportunities Identified for the Healing University ED" shows the improvement opportunities identified by the ED team.

Step 4: Design an Achievable Future State

In developing the ideal state, you were encouraged to act as though there were no constraints. Now it is time to be more practical. The future state map should depict what you believe can be accomplished in the next 6 to 12 months. Here are some key points to keep in mind:

▲ First, remember what you are trying to achieve. Identify improvement goals based on the outcomes measures you established at the beginning of the event (see step 4 in Chapter 12 on the current state).

CASE STUDY

Improvement Opportunities Identified for the Healing University ED

1. Improved processes
   a. Consider bedside testing for isolated critical care.
   b. Develop fast track with PA/NP to see/treat now.
   c. Combine greet and mini-registration processes.
   d. Combine greet and triage.
   e. Ensure quick triage on all patients.
   f. Do all registration at bedside.
2. Layout changes
   a. Join areas by a common, closer access.
   b. Triage and greeter layout and work assignments could be brought together.
   c. Close hall from main hospital to ER.
   d. Move lab printer away from the desk.
3. Signage and direction
   a. Install better signage at the entrance to ED and the parking area; add directional signage from garage to ED.
   b. Direct MRI parking to visitors' garage. Use valet parking for patients in ER.
   c. Improve access from main ED.
   d. Clarify campus signage with directions to ER or Information Center.
   e. Move security officer outside to control parking and direct as needed.
4. Elimination of paperwork
   a. Eliminate clipboards.
   b. Eliminate printed charts.
   c. Eliminate mini-registration form.
   d. Eliminate duplicate mini-registration form.

(continued on next page)

CASE STUDY (continued)

5. Changing roles
   a. Team works to push for more float staff and more RNs for sicker patient areas.
   b. Designate one person to be in charge and to monitor and direct the throughput. Be a problem solver, not a caregiver.
   c. Charge nurse should be director of and have oversight to operations.
6. Miscellaneous
   a. Physicians should standardize the process.
   b. Consistent registration process should be available 24/7.
   c. Identify protocol order sets that include lab orders to generate specimen labels.
   d. Transition orders should be expedited
   e. ED tests should be prioritized by the ED.
   f. Label blood specimens at bedside.
   g. Create a better prompt to technicians for lab draw; retrain for labeling.
   h. Consider the use of other protocols to begin care in triage.
7. Technology
   a. Identify issues that prevent physicians from using mobile devices and develop solutions (i.e., voice recognition, typing skills, larger laptops, etc.)
   b. Use phones for RN communication.
   c. Have IT fix computers in patient rooms.
   d. Use pager systems for orders.

▲ Apply Lean and healing pathway principles to the greatest extent possible, i.e., flow, removal of non-value-added activities, pull, and enhanced patient experience.
▲ Work from your current state and ideal state maps. Incorporate the ideas from your flow walk, i.e., eight healing impediments and eight wastes.

HINT

Don't let this step take too long. Some parts of the plan are not easy to see yet. Focus on what is easy to see, and plan to make an immediate and visible impact.

As you consider the desired future state, focus on areas where there is the greatest opportunity for impact:

▲ What are the short-term goals?
▲ Where is the obvious impact?
▲ Where is there immediate patient benefit?
▲ Where is there staff benefit?
▲ Where is there inventory and supply improvement?
▲ Where is the money?

This will allow you to create an achievable future state map. An example is provided in Figure 13.2.

Once the future state map is completed, the value-added time and non-value-added time should be determined and compared with the current state. See Table 13.2.

Step 5: Identify and Prioritize Opportunities

The improvement opportunities you have identified will have varying degrees of difficulty and impact. It is likely you have sorted and prioritized

Figure 13.2 Future state value stream map.

Table 13.2 Comparison of Current State and Future State

Performance Data	Current State	Future State
Total cycle time	258.9 minutes	119 minutes
Total value-added time	55.8 minutes	47.8 minutes
Total non-value-added time	203.1 minutes	71.2 minutes
Value-added percent	21.6	40

these opportunities in step 3. Now you will determine the best way to pursue each of the prioritized opportunities.

Recall the healing pathway transformation model from Chapter 3, reproduced as Figure 13.3. (Refer to Chapter 3 if you want a reminder of what each involves.)

The model shows that the opportunities identified during the healing pathway analysis will be pursued using one of three methods:

▲ Rapid improvement events (RIEs)
▲ Just Do Its (or JDIs)
▲ Projects

In this step you will sort these opportunities by impact and difficulty. A demand matrix is a useful tool for this exercise. The matrix should be drawn and placed in clear view, and the team will place the improvement opportunities on yellow sticky notes in the appropriate location. The typical layout for a demand matrix is shown in Figure 13.4.

Healing University's actual demand matrix is shown in Figure 13.5.

Summarize the opportunities into the appropriate categories as shown in Figure 13.6.

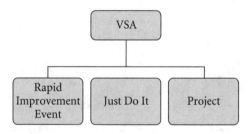

Figure 13.3 Healing pathway transformation model.

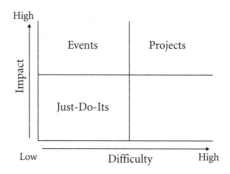

Figure 13.4   Typical layout for a demand matrix.

Step 6: Develop a Rapid Improvement Plan

The final output from the healing pathway analysis is the rapid improvement plan. This plan contains the rapid improvement events, the Just Do It events, and the projects identified during the event.

In step 5 the team organized the improvement opportunities into events, projects, and JDIs. During this step, begin by using yellow sticky

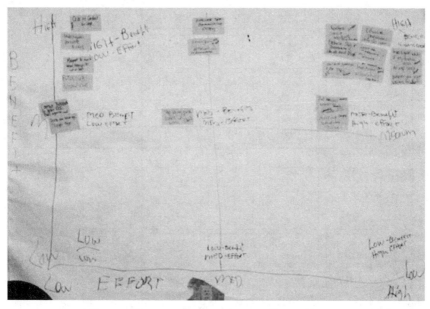

Figure 13.5   Healing University's demand matrix
for the ED healing pathway analysis.

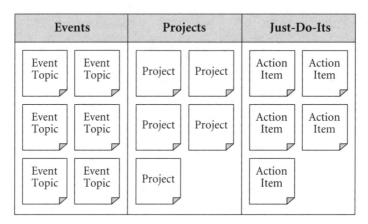

Events		Projects		Just-Do-Its	
Event Topic	Event Topic	Project	Project	Action Item	Action Item
Event Topic	Event Topic	Project	Project	Action Item	Action Item
Event Topic	Event Topic	Project		Action Item	

Figure 13.6 Sorted summary of improvement opportunities.

notes to denote each opportunity. Going again to the wall, start to assemble a time line for when they will be accomplished. Figure 13.7 depicts a graphical layout that can be used effectively in a team setting.

	May	Jun	Jul	Aug	Sep	Oct
Events	Event Topic	Event Topic	Event Topic	Event Topic		
	Event Topic	Event Topic	Event Topic	Event Topic		
	Event Topic	Event Topic	Event Topic			
Projects	Project ├———┤		Project ├————————┤			Project
		Project	Project ├————┤		Project ├————┤	
Just-Do-Its	Action Item	Action Item	Action Item	Action Item		
	Action Item	Action Item	Action Item			

Figure 13.7 Initial time line for a rapid improvement plan.

Formalize this schedule in a rapid improvement plan based on program milestones and need-by dates. Assign responsibilities for leadership of each activity. This rapid improvement plan will serve as the road map for the transformation of the healing pathway. Combined with a review of the desired outcome measure, the rapid improvement plan is also the tool that the executive sponsor and team leader will use to monitor progress. The sample form shown in Figure 13.8 can serve to document the final rapid improvement plan.

HINT

At the end of most healing pathway analyses, a wrap-up session is completed. It is here that the group shares the improvements or plans for improvements. It is meant to be a celebration and a reward for the team's hard work, as well as an opportunity for the senior leaders to be informed of the team's plans and progress. Senior management, the executive sponsor, and other interested parties should be invited. The team should take responsibility for presenting various portions of the out-brief. The out-brief should begin with a review of the charter and walk guests through the major steps of the analysis, ending with a review of the rapid improvement plan.

Event	Project	Just Do It	Description	Who	Plan Dates		Comments	Status
					Start	Finish		

Figure 13.8 Sample rapid improvement plan form.

Healing University's rapid improvement plan for the ED healing pathway analysis is shown in the following case study.

CASE STUDY

Healing University Rapid Improvement Plan for Emergency Department Healing Pathway Transformation

Event	Project	Just Do It	Description	Start	Finish	Comments	Status
				Plan Dates			
X			Access (Scope: from call ahead to triage decision)	6/22/12	6/28/12		
	X		Standardize orders				
X			Team care (Scope: from transfer to room to MD assessment)	6/22/12	6/28/12		
X			Rapid DX testing (Scope: from order entry to results)	6/22/12	6/28/12		
X			D/C and checkout	6/22/12	6/28/12		
	X		Transitional admission: standardized orders				
X			Bed ahead				
X			Decrease hall clutter (5S)	6/8/12	6/8/12	Increase safety	
		X	Staff follow dress code	6/8/12	6/8/12	Standards	
		X	Establish time standard for transports	6/8/12	6/15/12	Increase patient throughput	
		X	Reinforce understanding of capabilities of IBEX	6/15/12	6/20/12	Increase efficiency	
		X	Relocate laundry cart (5S)	6/10/12	6/20/12	Increase efficiency/safety	
		X	Reinforce charge nurse function to be responsible for monitoring/oversight of operations	6/10/12	6/20/12	Efficiency/throughput	
		X	Fix log on problem with bedside PCs	In Process		Increase efficiency	
	X		RNs use companion phone for communication	6/9/12	6/10/12	Time 5 min./Traffic decrease	
	X		Establish standard process for lab labels	6/15/12	7/15/12	Increase efficiency, decrease errors, increase patient safety	
	X		Close ED hall to thru traffic; corridor project	6/15/12	6/20/12	Decrease congestion	
X			Parking control (valet)	6/15/12	6/20/12	Increase customer satisfaction	
	X		Redirect ambulance walk-in traffic	6/10/12	6/15/12	Decrease traffic congestion	
	X		Signage study	6/10/12		Increased communication	

HINT

You may wish to capture the results of the healing pathway analysis in a report. This will serve to document and memorialize the team's work, and it is appropriate for sharing with interested individuals who did not participate in the event. An executive summary should be included; an example is provided in the next case study.

CASE STUDY

Executive Summary from the Healing University HPA Report

Background: A healing pathway analysis (HPA) of the emergency department (ED) process was conducted to improve patient safety and satisfaction, increase employee satisfaction, and improve the flow and throughput in the ED. During the HPA a cross-functional team representing most hospital functions searched for ways in which to streamline the ED process to obtain the above objectives.

Current State: The HPA team determined the current state utilizing a scenario in which a 35-year-old female presented with abdominal pain. Moving that patient through the ED healing pathway required a total cycle time of over 4 hours, with actual value-added time slightly less than an hour. The process contains a total wait/non-value-added time of more than 3 hours with approximately 22 percent value-added time (touch time). The team's objective is to drastically cut the total cycle time by reducing waiting and non-value-added activities.

Future State: The HPA team developed a desired future state and estimated the improvements in cycle time that could be achieved. In the future state developed by the team, the total cycle time was cut over 50 percent to 119 minutes for the emergency department process; non-value-added time was reduced 65 percent to 71 minutes; value-added time was reduced 14 percent to 48 minutes; and overall value-added time was improved to 40 percent.

Rapid Improvement Plan: The rapid improvement plan generated during the value stream analysis identified seven rapid improvement events (RIEs), six Just Do Its, and seven projects that are needed to create the future state. The first RIEs are scheduled for June 22 to 28.

Expected Benefit: The rapid improvement plan identified during the HPA should drastically cut the lead time that a patient spends waiting for treatment, while preserving the amount of physician-patient interaction that is a requisite part of the patient's experience. The vastly improved ED system should increase patient throughput while increasing safety and satisfaction.

CHAPTER 14

Ensure Healing Pathway Transformation Results

During the healing pathway analysis, a new future state was designed, and a rapid improvement plan was developed that identified the actions required to achieve it. Teams and their leaders often find these events to be very fun, exciting, and energizing. The challenge comes when team members leave the event and go back to the real world and their day-to-day jobs. It is sometimes difficult for the leader and the team to stay engaged, follow up, and bring the desired and exciting future state to reality. That is what this chapter is all about: *Checking* on the results and *Acting* to lock them in.

Like enterprise transformation results, healing pathway results can be measured in two ways: deployment and outcome. *Deployment* refers to execution of the rapid improvement plan. For example, are the Just Do It events, rapid improvement plans, and projects being executed as scheduled? *Outcome* here refers to the impact on key metrics of the healing pathway. Does the execution of the rapid improvement plan yield the planned results?

The executive sponsor, team leaders, and other key individuals should continue to meet on a regular and ongoing basis to review deployment and outcome results of the healing pathway transformation. Objectives of this ongoing leadership role are as follows:

▲ Demonstrate by their participation and leadership that transformation of the healing pathway is a priority.
▲ Monitor healing pathway transformation efforts and sustain results.
▲ Identify and remove barriers.
▲ Continue to reenergize the transformation.

▲ Reprioritize efforts and initiatives as appropriate.
▲ Standardize Best Practices.

What Do I Do?

▲ Develop a healing pathway dashboard.
▲ Meet biweekly to ensure execution.
   ▼ Review the dashboard.
   ▼ Review the rapid improvement plan.
   ▼ Eliminate barriers.
   ▼ Reprioritize as appropriate.
▲ Standardize Best Practices.

Step 1: Develop a Healing Pathway Dashboard

The dashboard is the tool the executive sponsor and transformation team should use to evaluate the outcomes of the healing pathway transformation initiative. It should be updated weekly and should include:

▲ Metrics that reflect the overall health of the healing pathway
▲ The specific metrics associated with transformation efforts

This dashboard is also the tool that the executive sponsor will use to brief the senior leadership team members during their regular reviews of the enterprise transformation. At the highest level, it should include metrics of interest to the senior leadership team. As an example, for the emergency department that might include average length of stay, patients left without being seen, and hours on diversion. For the revenue cycle, the dashboard might include average accounts receivable (A/R) days, days' cash on hand, and percent of self-pay. These metrics will be determined at the senior leadership team level and should cascade into the healing pathway dashboards as key determinants of the pathway transformation.

The dashboard should also include specific metrics associated with processes that make up the healing pathway. For example, if the team is responsible for transformation of the emergency department healing pathway, then the time to triage and average turnaround time for the lab may be important metrics for the team to review. Similarly, if the team is responsible for transformation of the revenue cycle, metrics such as average bill hold days, number of claim edits, and accounts to follow up may be critical to your effort. These key measures will be determined by the transformation team during the healing pathway analysis and will provide an objective assessment of the team's achievements. The following case study provides the dashboard for the Healing University ED project.

CASE STUDY

The ED transformation team developed the following dashboard to track results of their healing pathway transformation. They documented baseline and future state objectives, and they employed a simple red-yellow-green color coding scheme to highlight progress (shown here in shades of gray).

▲   Red (dark gray): Inadequate or insufficient progress
▲   Yellow (light gray): Improving but not at goal future state
▲   Green (medium gray): Future state performance achieved

Metric	Baseline	Future	Jan	Feb	Mar	Apr	May
Average LOS	4.84 hours	3.8 hours	4.7	4.3	4.3	4.1	3.8
LWBS	7.6 percent	5 percent	7.8	7.3	6.9	5.8	5.5
Diversion	94 hours	0	102	54	45	0	0
Productivity	4.4 FTE/UOS	4	4.2	4.4	3.8	4.2	4
Patient satisfaction	58th percentile	80th percentile	50	60	62	68	74
Time to triage	18 minutes	10 minutes	18	16	16	12	10
Lab results	56 minutes	30 minutes	55	36	28	34	29

Step 2: Meet Biweekly to Ensure Execution

This step is one of the most critical for ensuring sustainment of the healing pathway transformation effort. Continued participation of the executive sponsor and transformation team in these regular and ongoing meetings will serve to maintain the drumbeat of the transformation.

The executive sponsor and transformation team should meet regularly to maintain energy, focus, and progress toward achieving the future state.

Initially, meetings should be held biweekly at a minimum. Later, as the future state is achieved, the meeting frequency can be spaced further apart.

Each meeting should follow a standardized agenda, which should include the following items:

▲ Executive sponsor's welcome and remarks
▲ Review of charter and key objectives of the healing pathway transformation
▲ Review of actions from prior healing pathway transformation meetings
▲ Review of healing pathway dashboard
▲ Review of the rapid improvement plan
▲ Out-briefs by team leaders of Just Do Its, projects, and rapid improvement events, including:
  ▼ Status of all action items
  ▼ Key accomplishments to date
  ▼ Barriers to success
  ▼ Help needed from the transformation team
  ▼ Next steps
▲ Problem solving and elimination of barriers
▲ Evaluation and reprioritization of initiatives as appropriate
▲ Action item review

Ensuring the drumbeat of ongoing review, discussion, and problem solving will keep your team engaged and energized, and it will maximize the probability of success for transforming the healing pathway.

Step 3: Standardize Best Practices

Standardizing Best Practices at the healing pathway level means:

▲ Implementing the desired future state (Best Practice!)
▲ Developing and documenting new and revised procedures as appropriate
▲ Ensuring these new procedures are fully implemented

Coupled with the biweekly transformation team meetings, this is the *Check-Act* portion of the healing pathway transformation. During the transformation team meetings, information will be reviewed and actions identified. These actions must be implemented. This is done in four sequential steps:

▲ Check to ensure the new process is achieving the desired results.
▲ Act to document new policies, procedures, and work instructions as appropriate.
▲ Provide appropriate training and education to team members.
▲ Conduct internal process audits to identify training and/or compliance opportunities.

To ensure consistent and enduring implementation of processes improved during their healing pathway analysis, Healing University developed an internal audit process as described in the next case study.

CASE STUDY

Healing University ED Transformation Team Process Audit

The ED transformation team members met on a regular and ongoing basis to review implementation of the rapid development plan they developed for achieving a future state. They ensured that new policies and procedures were documented and deployed through training and integration into the documentation system.

After an apparently successful deployment, April's results for lab turnaround times slipped significantly:

Emergency Department Dashboard

Metric	Baseline	Future	Jan	Feb	Mar	Apr	May
Lab results	56 minutes	30 minutes	55	36	28	34	29

The team initiated a real-time process audit to be conducted over the next 3 days and designated the team leader of the lab's rapid improvement event to lead the audit. Upon review it was determined that new policies and procedures had been developed and deployed shortly after the event, as planned. This included training of all ED lab technicians.

(continued on next page)

CASE STUDY (continued)

However, the process audit revealed that not all ED technicians were labeling ED lab specimens as "Stat – ED," with the new labels the team had developed, in accordance with the new procedures. The audit team readily discovered that the root cause of the problem was a lack of new labels. Apparently only enough labels for an initial dry run were ordered, and when those were exhausted, the technicians did what they had to do—revert to using the labels that were available.

The new labels were added to the supply system and a kanban system was implemented to ensure their availability. Lab turnaround results improved immediately thereafter.

PART IV

Process Transformation

CHAPTER 15

Just Do Its

What Is It?

A Just Do It is one of the three types of actions that a transformation team may identify in a rapid improvement plan for transformation of a healing pathway. The other two are rapid improvement events, described in Chapter 16, and projects, described in Chapter 17. In Chapter 3, a Just Do It (JDI) was defined as follows:

> Just Do It (JDI)—An action for which no additional study or decision making is necessary. All information and knowledge necessary to decide on the process change is available, and the team readily and easily agrees. The team has decided it should be done, and all that is left is for someone to make it happen.

In contrast, a *rapid improvement event* (RIE) is a facilitated, structured, team-based event aimed at further refining and actually implementing improvements to portions of a value stream. A *project*, which includes both Lean and Six Sigma projects, typically entails additional analysis, is typically data analytic, and is often conducted by an engineer or small team to determine the best approach for achieving a desired outcome.

When Do I Use It?

The demand matrix described in Chapter 13 will assist you in determining when a Just Do It is the appropriate change tool. In addition to that, the defining characteristic of a Just Do It is described by its name. No additional study or analysis is required. The team has determined that it is one of the things that needs to be done. This is in contrast to RIEs and projects. For those, team members determine that there is an opportunity, but they do not currently have adequate information to determine the appropriate action. In these cases there is a need for either a team-based RIE or a more analytical project.

NOTE

Just Do Its are not necessarily easy or quick! For example, enlarging the ED might be identified as a JDI. What makes it a Just Do It is that it just needs to be done. But it will take time and money. That said, it turns out that most JDIs are relatively quick. Just keep in mind that they are not necessarily so.

An overview of the Just Do Its arising from the Healing University ED project is shown in the next case study.

CASE STUDY

Healing University Emergency Department Just Do Its

The rapid improvement plan developed during the ED healing pathway analysis yielded a total of six Just Do Its.
These tasks were included:

▲ Ensure that staff follows the dress code.
▲ Establish a time standard for transports.
▲ Reinforce the understanding of capabilities of the patient tracking system.
▲ Relocate the laundry cart.

(continued on next page)

CASE STUDY (continued)

Event	Project	Just Do It	Description	Plan Dates Start	Finish	Comments	Status
X			Access (Scope: from call ahead to triage decision)	6/22/12	6/28/12		
	X		Standardize orders				
X			Team care (Scope: from transfer to room to MD assessment)	6/22/12	6/28/12		
X			Rapid DX testing (Scope: from order entry to results)	6/22/12	6/28/12		
X			D/C and checkout	6/22/12	6/28/12		
		X	Transitional admission: standardized orders				
X			Bed ahead				
X			Decrease hall clutter (5S)	6/8/12	6/8/12	Increase safety	
		X	Staff follow dress code	6/8/12	6/8/12	Standards	
		X	Establish time standard for transports	6/8/12	6/15/12	Increase patient throughput	
		X	Reinforce understanding of capabilities of IBEX	6/15/12	6/20/12	Increase efficiency	
		X	Relocate laundry cart (5S)	6/10/12	6/20/12	Increase efficiency/safety	
		X	Reinforce charge nurse function to be responsible for monitoring/oversight of operations	6/10/12	6/20/12	Efficiency/throughput	
		X	Fix log on problem with bedside PCs	In Process		Increase efficiency	
	X		RNs use companion phone for communication	6/9/12	6/10/12	Time 5 min./Traffic decrease	
	X		Establish standard process for lab labels	6/15/12	7/15/12	Increase efficiency, decrease errors, increase patient safety	
	X		Close ED hall to thru traffic; corridor project	6/15/12	6/20/12	Decrease congestion	
X			Parking control (valet)	6/15/12	6/20/12	Increase customer satisfaction	
	X		Redirect ambulance walk-in traffic	6/10/12	6/15/12	Decrease traffic congestion	
	X		Signage study	6/10/12		Increased communication	

▲ Reinforce that the charge nurse is responsible for monitoring and oversight of operations.

▲ Fix log-on problems with bedside PCs.

Most were implemented immediately or within a few days. Fixing log-on problems with the COWs (computers on wheels) turned out to be a much more challenging problem than initially suspected. It turned out that the machines were outdated and could not properly interface with the system to provide bedside registration. Since bedside registration was critical to achieving the future state, the executive sponsor requested that IT put a priority on replacement COWs. These were procured and installed within two weeks!

A case study demonstrating innovation in physician office design at Healing University is shown in the following case study.

CASE STUDY

Innovation in Physician Office Design at Healing University

Background

Healing University Healthcare System owns and operates physician practices within their community. They conducted a Lean event as part of the design and planning for a new practice.

Objectives

During the healing pathway analysis, the team developed the following goals and objectives for the new physician practice:

- ▲ Less stress on physicians and staff
- ▲ Happy MDs
- ▲ Increase in market share
- ▲ Patient satisfaction at the 95th percentile
- ▲ Additional 9000 patients per year
- ▲ Increase in admssions
- ▲ Increase standard of care

Key Operational Principles

During the future state portion of the HPA, the team envisioned an ideal future state and derived the following principles. The intent was to maximize the implementation of these aspirational objectives in an achievable future state:

- ▲ Pull, smooth, one-way and soothing flow
- ▲ Efficient and effective checkout process
- ▲ Minimal waits
- ▲ Check-out done at check-in
- ▲ Minimal physician walking
- ▲ Nurse access for physicians' ease and utility
- ▲ Design for paperless process
- ▲ Central accessibility of lab and x-ray
- ▲ Aesthetically pleasing for the patients

CASE STUDY (continued)

Just Do It

The team developed a rapid improvement plan, which captured the activities and actions required to achieve the desired future state. One of those actions was a Just Do It to apply an innovative design to the entryway of the practice. The architect was given guidance to maximize the practice's ability to have an efficient and effective checkout process, minimizing the physical and temporal difference between the two. The innovative design of that area of the practice is shown below.

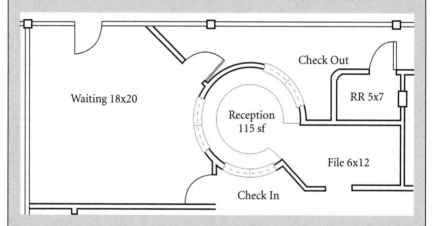

The design was ultimately realized in the new physician practice, and it greatly improved patient flow and satisfaction, while also decreasing staff workload and improving staff satisfaction.

CHAPTER 16

Rapid Improvement Events

What Is It?

A *rapid improvement event*, or RIE, is an accelerated team-based improvement event. It is results oriented. Unlike the healing pathway analysis, which is primarily a planning event, the objective of an RIE is the actual implementation of change. It involves a focused group of people who "do the work" and understand it. The rapid improvement event does not have a rigid structure but instead has multiple tools that may be used by the facilitator depending on the goal to be achieved. Rapid improvement events are also known as *kaizen events*, *kaizen blitzes*, or *kaikaku events*.

Rapid Improvement Event in a Nutshell

▲ **Purpose:** Problem identification, root cause analysis, new process design, and implementation
▲ **Who:** Cross-functional team including individuals involved in the process
▲ **Significant outcomes:** 50 to 75 percent implementation of the new design by end of the week and 100 percent implementation within 30 days
▲ **Length:** 1 to 5 days

What Does It Do?

Rapid improvement events by definition are used to reduce the time needed to benefit from Lean initiatives. The ultimate goal of an RIE is a new and improved process. Some changes will be implemented by the end of the event. Others will be captured on a rapid improvement newspaper with assignments and dates for near-term completion.

In today's world, when team members juggle multiple responsibilities, roles, priorities, and tasks, it can be challenging to find time to improve the process. At the same time, dealing with chronic process problems, rework, and other inefficiencies is a vicious cycle from which we have to help our people escape. Eliminating those time wasters will improve the lives of everyone. At the end of the day, it may be necessary to initially work even harder, i.e., engaging in process improvement while still accomplishing the work. Use of an RIE can make the additional process improvement effort manageable and can yield immediate results. A well-defined and well-facilitated rapid improvement event can provide significant improvements in a very short time.

Figures 16.1 and 16.2 contrast the traditional and RIE ("blitz") approaches to process improvement.

The underlying philosophy of an RIE is rapid change, as captured by the slogan "Do it now. Do it fast." It is more important to make quick but significant steps that approximate the desired future state now than to

Typical Project = 2–3 Months

Traditional Mindset ⟶ Analyze/Optimize/Final/Part Time/Do It Right the First Time

Figure 16.1 Traditional process improvement mindset.

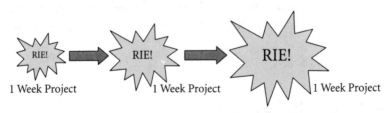

Repetitive RIEs . . . Improvement . . . Cycles of Learning

RIE Mindset ⟶ Better, Not Best/Just Do It/Do It Now, Do It Fast

Figure 16.2 Rapid improvement event (blitz) process improvement mindset.

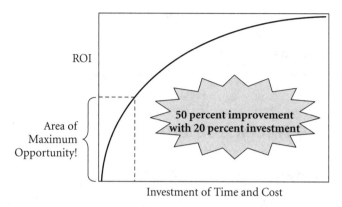

Figure 16.3   ROI greater at the front end of the improvement curve.

engage in a lengthy, and most likely unsuccessful, improvement effort aimed at achieving "future perfect."

As with many improvement efforts, the Pareto principle applies here as well: the return on investment (ROI) is greater at the front end of the improvement curve. This is illustrated in Figure 16-3. RIEs take advantage of the principle.

What Do I Do?

Rapid improvement events will have significant variations in content and structure. They are the foundational team-based problem-solving tool for achieving opportunities identified during the healing pathway analysis and documented in the rapid improvement plan. For example, they may be used to design and implement:

▲ Continuous flow
▲ Standard work
▲ Error proofing
▲ Quick changeover
▲ Visual management
▲ And any other number of team-based process improvements

The agenda will vary according to the specific objectives of the RIE. However, most RIEs will include the following phases:

- ▲ Development of the team charter
- ▲ Training of the team
- ▲ Study and implementation
- ▲ Development of a rapid improvement newspaper
- ▲ Follow-up

How Do I Do It?

Step 1: Develop the Team Charter

As with an HPA, a team charter should be developed to guide the team's efforts. The high-level objectives of the RIE were likely determined by the HPA team. Meet with the executive sponsor and team leader to complete and formalize the charter.

TIP
The executive sponsor will not be as involved in the RIE as in the HPA. This will be one of many for which he or she has responsibility. It is therefore all the more important that you have his or her guidance and commitment up front. In addition, invite the executive sponsor to the event out-briefs to stay in tune with the team's efforts.

To finalize the charter:

- ▲ Define the scope and target area of the RIE.
- ▲ Identify and clearly describe the RIE outcome objective(s).
- ▲ Propose and finalize the team membership and roles.
- ▲ Establish timing and scheduling requirements.
- ▲ Prepare other organizational resources (not on the team) impacted by the kaizen event for upcoming events.

During this meeting with the executive sponsor and team leader you will also want to begin work on the quad chart. A quad chart is a one-page summary of the event, and it is useful for tracking the team's accomplishments as well as telling the "HPA story," otherwise known as the

HINT

A cross-functional team will maximize the rapid improvement event's probability of success. Membership should include:

▲ People in the process
▲ People who supply process inputs
▲ People who are recipients of process outputs
▲ People well versed in Lean methods
▲ People who are completely unattached to the process but ask tough questions

Quality Improvement (QI) story, to others. As depicted in Figure 16.4, a quad chart has four sections:

▲ **Project description:** Expected outcome, executive sponsor, team leader
▲ **Key improvements and accomplishments:** Improvements made, key accomplishments during the RIE, major actions pending, and managing the process

I. Project Description:	II. Key Improvements and Accomplishments
Expected Outcome:	
Executive Sponsor:	
Team Leader:	

III. Healing Pathway Metrics/Benefits:	IV. Process Tracking Metrics

Metrics	Baseline	Future	Current

Metrics	Current	Future	By When

Status as of:

Figure 16.4   Sample quad chart.

▲ **Healing pathway metrics:** Previously determined by the executive leadership team and/or the HPA team

▲ **Process tracking metrics:** Measures the processes that the team expects to impact from this RIE

Before the rapid improvement event you will want to complete Sections I and III. Sections II and IV will be completed at the end of the event.

Step 2: Training

As with the healing pathway analysis event, team members should be trained at the beginning of the rapid improvement event on the appropriate tools and techniques to be used in the event. Do not assume that team members have the same understanding or the correct understanding of Lean principles or methods.

Consider these subjects as a minimum:

▲ Review of the charter
▲ Review of the quad chart
▲ Overview of a healing pathway analysis
▲ Actual results of the healing pathway analysis that stimulated this RIE
▲ Current state and future state mapping (enterprise, value stream, function level) as appropriate
▲ Process flow diagramming
▲ Continuous flow
▲ Lean metrics
▲ Team dynamics
▲ Team decision making

TIP
The amount of required training will vary depending on the experience of the team and the complexity of the event. In most cases a 4-hour introductory session will be sufficient. Additional training can be provided in real time as the team is engaged in problem-solving events during the rapid improvement event.

Step 3: Study and Implementation

The team will now study the current process in greater detail and will experiment with potential solutions. The tools that are used to identify the opportunities are similar to the those used in the healing pathway analysis. However, with an RIE, the team is "diving in deeper," and it will also make changes and experiment "on the fly." The tools that may be used include:

1. Spaghetti diagram
2. Process flow map
3. Inventory deployment map
4. 5S and visual management
5. Standard work
6. Fishbone diagram

The team will further identify and understand process wastes, impediments to healing, and obstacles to patient or process flow. This combination of thoughtful reflection and active problem solving will enable the team to envision the future state, explore impediments, and develop workarounds for those impediments.

The future state and current state maps will greatly assist in this effort. The gaps between the future and current states are what should be addressed by the RIE team. Prioritize what can be accomplished in a very short time. Remember, the objective is to "do it now, do it fast." However, don't discard those great ideas that are not readily or immediately implementable. They will provide fodder for future improvement activities.

Implement changes required to achieve the future state. In so doing, consider the impact on and requirements for the following five process elements:

▲ **People:** Define their roles and responsibilities.
▲ **Policies and procedures:** Define and document new work content and flows, how decisions will be made, what the new work standards are, and so on.
▲ **Data:** Define what information is required.
▲ **Information technology:** Define new information system technologies, work flows, rules, etc.
▲ **Facilities and equipment:** Design new layouts and acquire necessary equipment and tools.

Step 4: Develop a Rapid Improvement Newspaper

A rapid improvement newspaper is a useful tool for tracking changes that are being made. It also allows for easy sharing with the group and can serve as the basis of your periodic status review meetings. The newspaper contains the following:

1. Item number for tracking
2. Problem to be resolved
3. Action needed: The work that needs to be done to improve the problem
4. Responsibility: Who is responsible for completion of the task
5. Create date: The date the problem and/or idea was created
6. Complete date: The date when the action should be complete
7. Progress: Four-part progress measure
8. Salient comments

Figure 16.5 is a portion of a blank rapid improvement event newspaper.

Figure 16.6 shows the rapid improvement newspaper developed during one of Healing University's rapid improvement events for the ED.

Item	Problem to be Resolved	Action Needed	Resp.	Create Date	Complete Date	Prog.		Comments
1						1	2	
						4	3	
2						1	2	
						4	3	
3						1	2	
						4	3	
4						1	2	
						4	3	
5						1	2	
						4	3	

1 2 / 4 3	Team agrees that item needs to be addressed	1 2 / 4 3 Problem has been explained to person responsible
1 2 / 4 3	ECD for solution has been recorded on newspaper	1 2 / 4 3 Item completed

Figure 16.5 Portion of blank rapid improvement newspaper.

CASE STUDY

Healing University's Emergency Department Rapid Improvement Newspaper

The Healing University RIE team was broken into three subteams. Each developed a rapid improvement newspaper to identify and track planned improvement activities. Here is what the access team developed:

Team: Access

Item	Problem To Be Resolved	Action Needed	Resp.	Create Date	Complete Date	Prog.	Results/Savings (Please Quantify)
1	Control parking lot	a. Security officer placed in lot/work instructions b. Addition of booth	Steve	02/24/12	03/31/12	1 2 4 3	Increase parking availability for ED pts.
2	Ability to communicate from greeter to triage to charge nurse	a. Secure walkie talkies b. Investigate Vocera	Pamela	02/24/12	03/31/12	1 2 4 3	ED staff better prepared for pts.
3	Addition of 2.1 FTEs for redcoat role	a. Confirm need b. Request staff c. Determine busiest time/need d. Develop work instruction	Aaron	02/25/12	03/31/12	1 2 4 3	Control of visitors
4	Need for armed police officer vs. redcoat	a. Get security staff input b. Review c. ED Mgt.	Steve/ Pamela	02/25/12	03/31/12	1 2 4 3	Safety/guest relations for emp/pts/visitors
5	Triage area retrofitted	Add equipment	Tom	02/24/12	03/31/12	1 2 4 3	Quick triage for pts
6	Wheel chair ramp in need of repair (at end of sidewalk)	Ramp is rough and uneven	Steve	02/24/12	03/31/12	1 2 4 3	Better access
7	Supplies and meds lacking for fast track	Addition of pyxis cabinets	Ed	02/24/12	03/31/12	1 2 4 3	Improve fast track processing and service
8	Equip fast track room	a. Remove desks b. Add ENT chair c. Add medical gases d. Add otoscope etc.	Tom	02/24/12	03/31/12	1 2 4 3	Frees up beds in back for sicker pts.
9	Convert equipment room to treatment room	a. New location for safe b. New location for equip c. New medical equip secured	Steve/ Tom	02/24/12	03/31/12	1 2 4 3	Treatment room available in Triage area
10	Lack of interface for mini reg	Bidirectional interface	Kelly	02/24/12	03/31/12	1 2 4 3	No duplication— mini reg goes away

1 2 4 3	Team agrees that item needs to be addressed	1 2 4 3	Problem has been explained to person responsible
1 2 4 3	ECD for solution has been recorded on newspaper	1 2 4 3	Item completed

Figure 16.6   Healing University's emergency department rapid improvement newspaper.

Step 5: Evaluation and Follow-up

These are the *Check-Act* steps of the RIE process transformation. After a rapid improvement event, a standing meeting should be set up with the team, including the executive sponsor or at least the team leader. The purpose of these meetings is to check on the progress, eliminate barriers, and ensure successful follow-through of the team's improvement plan. The action items on the newspaper should be reviewed and updated and any new problems identified. This phase is critical to the success of the rapid improvement event.

On a regular basis, perhaps weekly at first and then occurring biweekly, the team should do the following:

▲ Use the rapid improvement newspaper as a tool to check the progress of actions to be completed.
▲ Use the quad chart to update process metrics and determine whether the event is meeting its objectives.
▲ Eliminate barriers.

Figure 16.7 describes and presents Healing University's completed ED quad chart.

HINT

You will want to capture real-time feedback on process changes. Accelerated process change techniques must be deployed to ensure that the new way "sticks." Here are some useful techniques:

▲ Create communication channels for associated feedback.
   ▼ Flip charts in the work area to record feedback.
   ▼ RIE blogs.
   ▼ RIE e-mail address where people can share their thoughts.
▲ Conduct formal surveys. Surveys should be generated for employee feedback and feedback from customers and/or suppliers, if impacted. Web-based survey tools make it easy to quickly gather input.

CASE STUDY

Healing University ED Quad Chart

Upon completion of the RIE, the Healing University team members documented their progress and identified process tracking metrics in a quad chart:

RIE: Healing U ED Front End

I. Project Description:	II. Key Improvements and Accomplishments
RIE of Emergency Department front end. **Expected Outcome:** Reduction in wait times, increase in value added time for patients, increased throughput capacity **Executive Sponsor:** Michael **Team Leader:** Tina	• Improvement newspaper—71 total items 13 complete • Healing pathways analysis—complete • Waste map—identified • Impediments to healing—identified • Future state—planned • Created standardized protocols • Change plan in place and being monitored • Process metrics established

III. Healing Pathway Metrics/Benefits:

Metrics	Baseline	Future	Current
Team member sat.	N/A	TBD	20%
Physician sat.	N/A	TBD	N/A
Patient sat.	72nd	75th	70th
Average LOS	4.62	3.0	4.37
LWBS	5.4	TBD	3.3
Hours on diversion			
Productivity			

IV. Process Tracking Metrics

Metrics	Current	Future	By when
Time to triage	17 min	3 min	3/31/12
ALOS	180 min	150 min	3/31/12
Avg door to room time	48 min	17 min	3/31/12
% pts in bed < 10 min	14%	65%	3/31/12

Status as of: 2/14/12

Figure 16.7   Healing University ED quad chart.

Lastly, document the team's efforts in a report or "QI story" presentation to share with the leadership team and other team members. An executive summary should also be developed, as it is useful for sharing results and stimulating interest and additional improvement activities across the organization. The following case study is an example of such a summary.

CASE STUDY

Healing University ED Rapid Improvement Event Executive Summary

Background: A healing pathway analysis (HPA) was conducted on the emergency department. During the HPA, the team identified a number of opportunities for improvement, including access, team care, and rapid diagnostic testing. These opportunities were the focus of this rapid improvement event (RIE). Subteams were established to focus on the three respective areas. This report provides some of the team's products, including the new process design of the ED, process improvements identified, the remaining tasks to be accomplished, and metrics for the ED.

Accomplishments: The RIE team was able to identify non-value-added activities that decreased patient quality of care, satisfaction, and throughput. A total of 71 improvement opportunities were identified, and 13 were completed during the event. Although actual data will be collected over the coming weeks, a simulation conducted during the RIE yielded times of 46, 50, and 19 minutes from patient entry until diagnostic testing results are available.

▲ The access subteam addressed issues relating to access to the emergency department. By improving the parking lot, security, and communication, the patient's arrival at the ER is much faster and simpler and is integrated into the treatment that occurs inside.

▲ The team care subteam focused on improving the triage and movement of a patient through the ED. By using parallel processing, the patient is now registered and triaged at the same time and is moved to a bed within a matter of minutes following arrival.

▲ The rapid diagnostic subteam worked on improving the quality and turnaround time of lab specimens and results. By adding a new rapid diagnostic technician (RDT), responsible for lab draws and transportation, they greatly improved first-time quality of lab specimens while allowing other care providers to focus on other

CASE STUDY (continued)

aspects of patient care. In addition, the RDT allows an increase in X-ray capacity.

Return on Event: Through the efforts of this event, additional capacity in three different areas: through the front end of the ED, the lab, and X-ray. The team estimated that ED capacity could increase by 50 percent. Data will be collected over the upcoming weeks and months to quantify exact improvements. The team concluded that the most significant benefit stemming from the RIE was the improvement of "broken" processes.

CHAPTER 17

Project

What Is It?

A *project* is one of the three types of actions that a transformation team identifies in a rapid improvement plan to achieve transformation of a healing pathway. The other two are Just Do Its, described in Chapter 15, and rapid improvement events, described in Chapter 16. As it is defined in Chapter 3:

> A *project* is an improvement effort that is typically data analytic and is often conducted by an engineer or small team to determine the best approach for achieving a desired outcome.

In contrast, a Just Do It (JDI) is an action for which no additional study or decision making is necessary. The team has decided it should be done, and all that is left is for someone to make it happen. A rapid improvement event (RIE) is a facilitated, structured, team-based event aimed at further refining and actually implementing improvements to portions of a value stream.

The term *project* is intentionally broad and includes improvement activities taken from both the Six Sigma and Lean tool kits. These include:

- ▲ Six Sigma projects
- ▲ Visual management
- ▲ Mistake proofing
- ▲ Kanban systems
- ▲ Setup reduction

Note that these tools and activities are not covered in great detail in this chapter. Summaries are included so that the reader has a complete guide to Lean healthcare. More detailed information on how to use these tools is included in other publications, including *The Lean Memory Jogger for Healthcare* (R. MacInnes & M. Dean. Goal/QPC, 2012).

Six Sigma Projects

What Is Six Sigma?

Six Sigma is a statistical concept that refers to the amount of variation in a process. More specifically, processes with Six Sigma variation have less than 3.4 defects per million opportunities.

In addition to being a statistical measure of variation, Six Sigma refers to a continual improvement methodology first developed by Motorola. The objectives of the Six Sigma initiative were to reduce defects and decrease variation.

What Is Lean Sigma?

Six Sigma can be considered an approach for reducing variation in processes. Lean is an approach for increasing process throughput by eliminating waste. These approaches complement each other well. On one hand, Lean lends itself readily to what might be considered "lower-hanging fruit." It is also often more intuitive, and its opportunities are often more readily perceivable to a group of individuals engaged in a Lean process such as a healing pathway analysis or kaizen event.

The Six Sigma element, on the other hand, is less intuitive and more analytically based. It yields opportunities through the application of more rigorous measurement and analysis. The reader will recognize that the approach advised in this book is an integrated model and a tool kit for achieving enterprise, healing pathway, and process transformation. The relationship between Lean and Six Sigma and their collective impact on the customer are depicted in Figure 17.1.

What Is DMAIC?

DMAIC is the foundational methodology used by Six Sigma to achieve process improvement. The DMAIC method involves five steps:

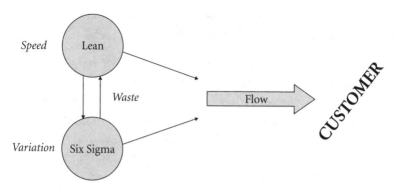

Figure 17.1   Relationship between Lean and Six Sigma.

▲ Define the project.
▲ Measure the current state.
▲ Analyze data to identify opportunities for improvement.
▲ Improve by testing proposed solutions and evaluating results.
▲ Control the gains by standardizing Best Practices.

The DMAIC model is very similar to the PDCA cycle. However, it emphasizes a greater reliance on measurement and analysis. While these steps are implicit in the Plan portion of the PDCA cycle, the DMAIC model makes these steps explicit. Thus it is considered a more precise model for process improvement for data-analytic improvement opportunities.

The next case study provides discussion and some initial analytic results from Healing University's revenue cycle project.

CASE STUDY

Healing University Six Sigma Project:
The Revenue Cycle Project

During the healing pathway analysis of the revenue cycle, the transformation team learned from some of the revenue cycle team members that there appeared to be many inappropriate "edits" identified by the scrubber. Those familiar with the revenue cycle will

(continued on next page)

CASE STUDY *(continued)*

know that the scrubber reviews claims for errors and identifies them as edits. The transformation team identified this as an opportunity for improvement, included it on the rapid improvement plan, and determined that a Six Sigma project was the most appropriate approach for identifying and eliminating the root cause. A Six Sigma black belt was assigned the project to "Review scrubber edits and eliminate as appropriate."

Initial analysis revealed that less than 50 percent of the patient claims clear the scrubber the first time through. Of those:

▲ 80 percent can be cleared by billing

▲ 20 percent are sent to HIM. Some can be resolved by HIM while others (CCI edits) are sent by HIM to the clinical areas. Time frame to clear these edits is 2 to 30 days.

Using the 80–20 rule to focus first on the highest-yield opportunities, further analysis was done to stratify the edits that can be cleared by billing. Pareto analysis yielded the following results:

204 Errors in 11 Edit Codes

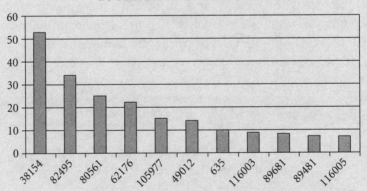

Guided by this analysis, the team collected additional data on the top error codes. The analysis confirmed team member suspicions that the scrubber was identifying inappropriate edits. In addition, some of the edits were unknowingly being reviewed twice. The vendor was

CASE STUDY (continued)

contacted to remove the unwanted edits, and the process was redesigned to preclude double edits.

Combined with successful completion of the other improvement opportunities identified in the rapid improvement plan, the revenue cycle transformation team achieved very impressive results.

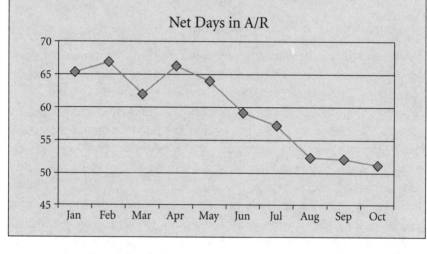

Net Days in A/R

The next case study provides a summary of an improvement project for Healing University's long-term care facility.

CASE STUDY

Skilled Nursing Facility Bed Management In-Sourcing Project

Background

Healing University Healthcare System owns and operates a large skilled nursing facility (SNF) within their community. For many years, they had contracted with one of the large bed vendors to repair, maintain, and

(continued on next page)

CASE STUDY (continued)

transport beds throughout the SNF. The contract had been in place for many years, with an annual cost to Healing University of over $357,000 per year. Bed vendor employees also tracked and transported the beds to patients, supposedly within a 1-hour window. The purpose of this project was to see if more cost-effective and timely bed management could be achieved.

In-sourcing as the Preferred Option

A team was developed that included nursing, material control, environmental services, and transport to look into the process and determine if they could more efficiently move the beds within the hospital. Over the course of the project, the team concluded that the vendor's response time to get a bed to the patients was acceptable. However, the team also concluded that Healing University could in-source the effort, meet or exceed current performance standards, and do so for a lot less money.

Healing pathway analysis yielded a number of opportunities for improvement. One of the challenges going forward was the SNF's ability to track and find beds. This was classified as a project and assigned to the facilities engineer to evaluate and implement. As a result of this project, a database was developed to track all the beds, including their current location and status as well as their functional status, i.e., whether repair was needed.

To implement, the engineer set up the database; then the team clearly labeled every bed in the SNF so it could be readily identified and set up a "parking lot" for where each bed was to be kept when not in use by a patient.

Results

This project yielded a recurring annual savings of $357,000, as the in-sourcing effort was taken on by the environmental services group without the addition of any internal FTEs. The beds are audited daily to ensure they are in their assigned locations, either on the floor or in their parking space. Response times continue to meet expectations.

Visual Management

What Is Visual Management?

Visual management is a set of techniques designed to improve workplace organization and standardization. The essence of visual control is that anyone entering the workplace, even those unfamiliar with the process, can very quickly and obviously see what is supposed to be happening, and what is under control and what is not. It is a way of organizing intelligence into the workplace to enable efficient and error-free performance.

Visual management is sometimes referred to as *5S visual management* or more simply as just *5S*. It includes the following five steps:

▲ **Sort:** Distinguish between what is needed and is not needed.
▲ **Set in order:** There should be a place for everything and everything in its place.
▲ **Shine:** Clean and look for ways to keep it clean.
▲ **Standardize:** Make standards obvious and maintain them.
▲ **Sustain:** Follow the new rules and procedures and follow up.

What Does It Do?

5S provides a basis for being a world class healthcare organization and a foundation for a disciplined and reliable approach to workplace and work flow management. It is also one of the most visible improvement activities and can stimulate excitement and commitment toward additional improvements efforts. Benefits include the following:

▲ Improved patient care
▲ Reduced cost of care
▲ A cleaner and more organized clinical area
▲ Reduction in non-value-added time
▲ Effective work practices
▲ Safer working conditions
▲ Increased job satisfaction

> ### TIP
>
> *The Lean Memory Jogger for Healthcare* is an excellent resource for guidance on conducting 5S events.

How Do I Do It?

Details on how to conduct a 5S are beyond the scope of this book but can be found in numerous other publications including *The Lean Memory Jogger for Healthcare*. However, the following keys to success may be useful in guiding your 5S implementation efforts:

▲ Get everyone in the workspace involved.
▲ Have leaders set expectations and conduct follow-up.
▲ Agree on a vision.
▲ Define roles.
▲ Start with a cross-functional team.
▲ Agree on and establish standards.
▲ Establish routines and checklists.
▲ Keep it simple.

Table 17.1 is a great resource for evaluating your organization's achievement in fully implementing 5S.

Mistake Proofing

What Is Mistake Proofing?

Mistake proofing is the use of process or design features to prevent errors or the negative impact of errors. Mistake proofing is also known as *poka-yoke*, a Japanese term for "avoiding inadvertent errors."

What Does It Do?

Human error is the largest source of mistakes in the real world. People are, after all, human. They make mistakes. Not surprisingly, we have historically focused error prevention on the human. Approaches have included

Table 17.1 5S Levels of Achievement

	Sort	Set in Order	Shine	Standardize	Sustain
Level 1: Just beginning	Needed and not needed items are identified; those not needed are gone	Items are randomly placed throughout the workplace	Key area items checked are not identified and are unmarked	Work area methods are not always followed and are not documented	Work area checks are randomly done and there is no 5S measurement
Level 2: Focus on basics	Necessary and unnecessary items are identified; those not needed are gone	Needed items are safely stored and organized according to usage frequency	Key area items are marked to check and required level of performance noted	Work group has documented area arrangement and controls	Initial 5S level is established and is posted in the area
Level 3: Make it visual	Initial cleaning is done and mess sources are known and corrected	Needed items are outlined, dedicated locations are labeled in planned quantities	Visual controls and indicators are set and marked for work area	Agreements on labeling, quantities, and controls are documented	Work group is routinely checking area to maintain 5S agreements
Level 4: Focus on reliability	Cleaning schedules and responsibilities are documented and followed	Minimal needed items arranged in manner based on retrieval frequency	Work area cleaning, inspection, and supply restocking done daily	Proven methods for area arrangement and practices are used in the area	Sources, frequency of problems are noted with root cause and corrective action
Level 5: Continuous improvement	Cleanliness problem areas are identified and mess prevention actions are in place	Needed items can be retrieved in 30 seconds with minimum steps	Potential problems are identified and countermeasures documented	Proven methods for area arrangement and practices are share and used	Root causes are eliminated and improvement actions include prevention

---

TIP

The Agency for Healthcare Research and Quality (AHRQ) has published an excellent resource on the application of mistake proofing in healthcare. The guide, entitled *Mistake-Proofing the Design of Health Care Processes*, is in the public domain and available online at no charge. This section is supplemented by excerpts from this excellent resource.

---

reprimands, retraining, and motivational talks to "be more careful" and "pay attention."

Mistake proofing is the science of preventing errors. Shigeo Shingo formalized mistake proofing as part of his contribution to the production system for Toyota automobiles. He determined that process design was a better way to prevent mistakes than our traditional human-focused approaches. Very simply, the approach is to improve the process to remove opportunities for error, or to make wrong action more difficult. When the opportunity for error cannot be removed, make it easier to discover the errors that do occur.

Poka-yoke devices fall into two major categories:

▲ Prevention
▲ Detection

In *prevention approaches*, the process is engineered so that it is impossible to make a mistake at all. An example is a sharps container for preventing injury or infections from used syringes. The design of the sharps container ensures that used sharps (needles) cannot be reached and removed after they are deposited. The used sharp is placed on the back part of the lid, which is then lifted to dump the sharp into the container. The lid's design makes it impossible to access the used sharps.

Another example in healthcare is the defibrillator. As a safety mechanism, all defibrillators require the activation of two separate buttons, held by one operator, to trigger discharge of an electric current across the thorax of patients in ventricular fibrillation. The two-button feature reduces the risk of accidental discharge or misplaced electrical shock.

TIPS FOR POKA-YOKE

▲ Eliminate unnecessary or duplicate steps.
▲ Use flowcharts to visualize the process.
▲ Keep it simple.
▲ Use many poka-yoke devices in a process.

When prevention is not possible, *detection approaches* signal the user when a mistake has been made, so that the user can quickly correct the problem. A simple example of this approach is the case where your car beeps if you leave the keys in the ignition.

When Should I Use It?

Mistake-proofing initiatives can be used to improve patient safety, eliminate workplace injury, and reduce failure costs. They are a robust design criterion for equipment and device manufacturers, but integration into hospital processes is still a huge opportunity. The referenced AHRQ report proposes the following four approaches for mistake proofing:

1. Mistake prevention in the work environment
2. Mistake detection (Shingo's informative inspection)
3. Mistake prevention (Shingo's source inspection)
4. Prevention of the influence of mistakes

Additional guidance is beyond the scope of this text, but the referenced AHRQ report and *The Lean Memory Jogger for Healthcare* are excellent resources for additional information.

Kanban Systems

What Is a Kanban System?

A kanban system is a method of using cards or other signaling devices as visual signals for triggering or controlling the flow of materials and supplies. Kanban is a Japanese term that essentially means signboard—a signal of consumption or need that triggers production or delivery of goods or services. Kanbans are essential to, and go hand in hand with, a just-in-time (JIT) operation and complement the push versus pull concept characteristic of Lean. In healthcare, one of the greatest applications of kanbans is in the supply chain.

There are two basic types of kanban signals:

▲ **Production kanban:** Signal that paces the flow of product through the value stream or healing pathway
▲ **Withdrawal kanban:** Signal that indicates the need to resupply materials consumed

Kanban signaling can be done in many ways:

▲ Physical observation of the number of items available
  ▼ Full = stop (maximum limit)
  ▼ Not full = run (below maximum)
▲ Computerized or electronic view that shows the number of available items and can provide move instructions
  ▼ Red = stop
  ▼ Green = priority run
  ▼ Black = OK to run
▲ Representative
  ▼ Cards, containers, carts, etc.

Why Use It?

In an ideal product or service delivery process, the need for materials and supplies would occur at a constant, predictable rate. In such an environment, an organization could easily predict usage requirements and provide exactly what was needed, when it was needed. There would never be stock outages or shortages, and there would be no excess inventory. But of course healthcare is not that type of environment. Demand is often unpredictable. Consider, for example, the rate of patients coming in to the emergency room. Kanban systems enable operations to flow smoothly while preventing overproduction and excess inventory. This is called providing a *just-in-time* (JIT) inventory.

How Do I Do It?

Lean enterprises often use what is called a *supermarket system* to achieve JIT inventory. The inventory on supermarket shelves is impacted by two interdependent processes: production and withdrawal (supply and demand). Customers withdraw what they want from the shelves. The store restocks depleted products. Kanban signaling indicates the need for either additional production or the supply of inventory for withdrawal. This supermarket concept can be applied in many healthcare processes. See Figure 17.2.

Details for how to set up a kanban system are beyond the scope of this book and can be found in numerous other publications including *The Lean Memory Jogger for Healthcare.* However, when you are designing a kanban consider these general design rules:

▲ Processes should be standardized and stabilized before implementation.
▲ Ensure quality prior to movement (jidoka).
▲ Produce or move nothing without a kanban.
▲ Consumers "pull" from the supplier.
▲ Suppliers should produce only to the kanban limit.

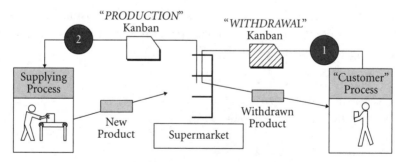

Figure 17.2   Supermarket system for just-in-time delivery.

Setup Reduction

What Is Setup Reduction?

Setup reduction is a method of analyzing your hospital processes to reduce the materials, skilled resources, and time required for setup or turnover. Setup reduction is also called *quick changeover*. The setup time is what is targeted for reduction. Setup time is the time required to do all activities required to change over from the end of one activity to the start of another activity. It will help your hospital teams reduce downtime and increase patient throughput.

Why Use It?

There are a number of advantages to addressing setup times in healthcare. These include the following:

▲ Team members can respond more quickly to changes in demand.
▲ Hospital capacity can be increased.
▲ Errors are reduced.
▲ Lead times are shortened.

Where Does It Apply in Healthcare?

Reducing setup time is a desirable outcome in many hospital processes, particularly those that impact the flow of patients into and out of the system and through their healing pathways. These include:

▲ In-patient room turnover
▲ Emergency department bed turnover
▲ Operating room turnover and setup
▲ Radiology setup
▲ Lab turnover and setup

How Do I Do It?

There are six steps to performing a setup reduction analysis.

Step 1: Document the Current State

▲ Make a video of the current state process. This will help identify repetitive and unnecessary movements. It also allows those who normally do the work to watch themselves.

▲ Record *all* steps required to perform the changeover. This will help identify unnecessary and inefficient steps.

Step 2: Separate External and Internal Work

▲ External work is work that can be done while the process runs, for example, work that can be done to prepare for the next surgical case while the current case is underway.

▲ Internal work is work that can be performed only when the process stops.

Step 3: Classify Setup Steps   Each setup step must be placed into one of the following classifications:

▲ **Preparation:** Activities that involve the gathering of tools, materials, or fixtures used in setups

▲ **Replacement:** Activities or tasks that involve the exchange or the connection and disconnection of tools, materials, or fixtures

▲ **Location:** Activities that involve the correct positioning of materials or equipment

▲ **Adjustment:** Activities involving actions repeated to obtain the first good outcome

▲ **Storage:** Activities that involve the placing of setup equipment and supplies in order after use

▲ **Other**

Use the changeover analysis work sheet shown in Figure 17.3 as a guide.

Step 4: Modify

▲ Examine each step and determine what can be improved or even removed.

▲ Create a spaghetti diagram (movement diagram).

▲ Document changes for use in the implementation stage.

Date:			Area of Hospital				Staff Involved	
From:	To:					Measurements by:		
#:	Work element		Running time (minutes/seconds)	Element time (minutes/seconds)	Category			
	Start		0:00:00		Internal Time	External Time	Improvement Ideas	
1								
2								
3								
4								
5								
6								
7								
8								
9								
10								
11								
12								
13								
14								
15								
16								
17								
18								
19								
20								

Figure 17.3 Changeover analysis worksheet.

Utilize the following modification techniques:

▲ Workspace organization (5S)
▲ Removal of external processes
▲ Improvement of internal processes (convert to external if possible)
▲ Standardization

Step 5: Implement Ideas

▲ Assign responsibilities to investigate and implement ideas
▲ Implement ideas (start with short-term ideas).
▲ Reevaluate the setup and document new setup time.
▲ Establish written procedures for the modified setup.
▲ Document time savings.
▲ Train others to perform the standardized procedure.

Step 6: Continue to Improve

▲ Establish metrics to measure improvement.

▲ Post the setup time at each location.

▲ Post the process performance charts.

▲ Do it again!

PART V

Continual Strategic and Operational Improvement

CHAPTER 18

Use Lean to Execute Strategic and Emerging Operational Objectives

The Journey Is Only Beginning!

The transformation process described in this book is a structured organizational approach for achieving organizational transformation. It has provided a road map for achieving the goals of the healing enterprise, specifically, excellent patient care provided by competent and caring staff, nurses, and physicians, with economically sustainable efficiency.

But now that you have successfully transformed your organization's people and processes, resulting in excellent operational performance, is that the end? Of course not! It is not the end unless you live and work in an environment of absolute stability. Healthcare is certainly not that. We are in perhaps the most dynamic time in the history of our industry, and continued operational excellence will require continued adaptation and improvement to meet the exigencies of the emerging marketplace.

So what should you do now? Well, here is the bonus! The process and infrastructure that you used to achieve organizational transformation can now be adapted as a process for strategically targeting improvement opportunities and aligning organizational efforts going forward. To recap, the transformation model provided in this book is comprised of three cascading and interdependent cycles, as fully described in Chapter 2:

▲ *Enterprise transformation* focuses on the healthcare services and enterprisewide systems, while directing specific healing pathway and value stream transformations.

▲ *Healing pathway transformation* focuses on high-level processes that constitute the healing pathways and value streams.
▲ *Process transformation* focuses on the specific processes that combine to form the healing and business pathways.

This same three-tier deployment hierarchy can be used as the standard to achieve strategic and operational improvement. It provided the appropriate infrastructure and authority cycle for the identification, prioritization, and execution of any initiative identified by the organization.

In its simplest form, the entire organizational transformation comprised the following steps, also initially described in Chapter 2:

▲ *Plan* the lean enterprise transformation.
▲ *Do* the enterprise transformation.
▲ *Check* the enterprise transformation.
▲ *Act* to eliminate barriers to lean enterprise transformation.

These same fundamental steps, coupled with the aforementioned deployment hierarchy, can be used to successfully implement strategic and operational objectives. The model was developed as a collaborative effort of a team of experienced, high-level healthcare executives and improvement practitioners. It has been playfully dubbed the "Legs" model, suggesting that it is an approach for "giving legs to" an organization's strategic and operational improvement objectives. Below is an acknowledgment of the contributing members of the collaborative effort.

ACKNOWLEDGMENT

This improvement targeting model was developed by the National Performance Management Team of healthcare provider Catholic Health Initiatives (CHI) and the author:

Mr. Bob Strickland, VP Operations Improvement, CHI
Mr. Mike McIntosh, VP Change leadership, CHI
Mr. Bob Cook, VP Change Leadership, CHI
Dr. Mark Dean, VP Performance Improvement,
Jewish Hospital and St. Mary's HealthCare

Strategic and Operational Alignment and Execution with Legs

What the Model Is

▲ It is a strategic improvement targeting model. It takes the annual objectives and metrics and targets specific, high-leverage improvement projects (specifically Performance Improvement based or not).

▲ It provides improvement activity alignment with the strategic plan.

▲ It is a method both of deploying the strategic plan alongside other identified operational objectives and of deliberately identifying improvement projects to attain the objectives as operationally defined by the metrics.

▲ It is a method for sustaining efforts through to completion.

What the Model Is Not

▲ It is not a strategic planning model. This model uses the output of the strategic plan at the level of annual objectives and metrics.

▲ It is not an organizational process improvement model. In other words, its objective is not continual process improvement per se. Rather, its objective is effective and successful implementation of organizationally defined priorities, both strategic and operational.

Problems with the Traditional Model for Targeting and Implementing Strategic and Operational Improvement Opportunities

Most organizations have a strategic planning process that yields organizationally critical objectives, goals, and plans. Then, most typically, there are also other, independent improvement efforts, identified either randomly or in response to the current "fires" of the day. This typical, relatively independent process is depicted in Figure 18.1.

While there is some overlap between the targeted opportunities, they are largely considered totally independent processes. There are a number of problems with this approach:

▲ It dilutes focus by propagating disparate organizational initiatives.

▲ There are often too many initiatives.

Figure 18.1 Relative independence of strategic and other improvement efforts.

▲ There is no centralized, comprehensive organizational prioritization of efforts.
▲ People are busy; the organization loses focus.
▲ Only some strategic goals and objectives are accomplished.
▲ Only some fires are put out and only some problems are solved.

Organizations will benefit by purposefully planning *all* the organizations' improvement efforts and putting in place an infrastructure to ensure their completion. This integrated systems approach is illustrated in Figure 18.2.

The Model at a High Level

A high-level depiction of the model is shown in Figure 18.3. A very detailed process and structure will be identified later in this chapter, but for now the key elements of the model are as follows:

Understand Strategic Plan, Goals, and Annual Objectives. One of the purposes of this model is to assist with the execution of the organization's strategic opportunities. Thus, a thorough understanding of these is a must.

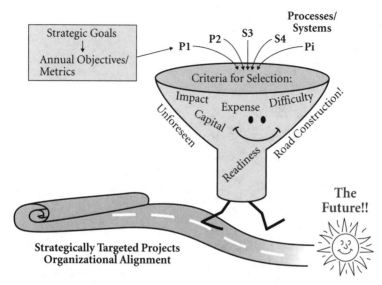

Figure 18.2 Integrated systems approach for prioritizing and aligning organizational objectives.

Identify Driving Metrics for Key Business Systems.   One of the unique aspects of this model is that it focuses directly on the attainment of clearly identified outcome metrics. Those metrics can be related to strategic objectives, for example, "Grow market share from 12 to 18 percent," and operational objectives, for example, "Increase patient satisfaction from the 54th to the 83rd percentile." Improvement of these key organizational metrics is the objective of this model.

Choose Systems and Processes to Improve.   The organization will methodically prioritize those opportunities that it believes will lead to movement of the key outcome measures. In other words, it will target for improvement those processes that directly impact the outcome measure. This will be accomplished through a largely participative process that ensures the contribution of appropriate stakeholders and knowledge.

Develop a Plan for Improvement.   Having identified the high-leverage processes that will move the desired outcome metrics, the leadership team will then develop a plan for improvement. Similar to the Lean transformation model described in this book, this will be a cascading plan, with

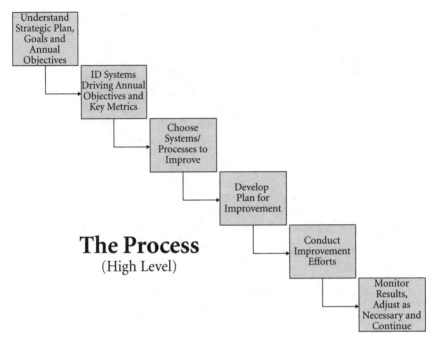

Figure 18.3   A high-level depiction of the strategically targeted legs model.

responsibilities assigned at the appropriate levels of the organization: enterprise, healing pathway, and process.

Conduct Improvement Efforts.   The same methodology that was used to deploy Lean transformation will serve as the set of tools for achieving ongoing improvement efforts.

Monitor Results, Adjust as Necessary, and Continue.   The reader will recognize these as the Check and Act steps used to deploy and sustain organizational transformation efforts. The same process for review and elimination of barriers will serve you well here.

Organizing for Improvement: Infrastructure and Roles

You will be well served to clarify the roles and responsibilities to ensure continued success of your strategic and operational objectives. Many of the roles and responsibilities will be quite similar to those required for the Lean transformation, with certain clarifications and slightly different responsibilities as highlighted below.

CEO

▲ Leads strategic planning
▲ Drives change and improvement
▲ Sets expectations and holds leaders accountable
▲ Creates an environment conducive to improvement and change

Senior Leadership Team

▲ Sets expectations for middle management regarding performance improvement and holds managers accountable
▲ Leads interdepartmental planning to achieve goals
▲ Monitors progress toward those goals
▲ Charters and monitors the progress of key functional and cross-functional improvement projects
▲ Removes barriers to progress

VP of Performance Improvement

▲ Facilitates the senior leadership team and works with senior leadership to translate strategic and annual business plans into performance improvement projects
▲ Provides technical data analysis, process improvement, and change leadership expertise in working with leaders and managers to identify, implement, and monitor the success of improvement ideas
▲ Facilitates key strategic process improvement projects

Teams.   Teams are organized as necessary. Not all improvement projects require a formal team. Membership is based on knowledge of the process and key stakeholders. There are two basic types:

▲ **Departmental:** These are teams who are working on a process or project that has boundaries within one department.
▲ **Cross-departmental:** These are teams who are working on more complex processes or projects that cross two or more departments.

Executive Sponsors

▲ Have authority to approve and disapprove process redesign changes.
▲ Are accountable for team progress and barrier removal.

Team Leader

▲ Typically, this is the process owner.
▲ The team leader is responsible for moving the team through its agendas and plans.
▲ The team leader is responsible for team logistics.
▲ The team leader is the key interface with the sponsor.

The Process in Detail

A detailed map of the transformation process is provided in Figure 18.4. Following is a detailed description of each of these steps and …

▲ Its purpose
▲ Who should lead it
▲ The major activities involved
▲ Potential tools to be used
▲ Some potential pitfalls to look out for

The numbers identified in the process steps below correspond to those depicted in the model.

Step 0: Preplanning Meeting

Purpose: To build a common understanding of this model and to assess readiness to proceed with application of the model.

Who Leads: VP of performance improvement or outside facilitator.

Major Activities:

▲ Review the model and discuss.
▲ Conduct an initial review of strategic plan and annual objectives:
  ▼ Are the objectives well aligned with the strategies?
  ▼ Are there good metrics, well aligned with the objectives, for monitoring progress on meeting the objectives?

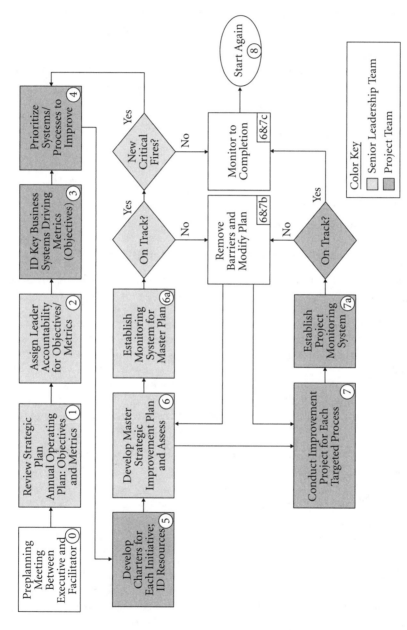

Figure 18.4 The "Legs" model for strategically targeting improvement opportunities.

▼ Is the number of objectives manageable given the resources available?

▼ Are there adequate process improvement (PI) resources available to guide the improvement of projects? (Rule of thumb: One trained facilitator for each process improvement project. Full-time facilitator can handle three to four projects [depending on complexity]; half-time facilitator, two projects, quarter-time facilitator, one project.)

Potential Tools: Discussion, presentation.

Potential Pitfalls:

▲ Misaligned metrics
▲ Poorly defined objectives
▲ Inadequate resources, too many objectives

Step 1: Review the Strategic Plan and Annual Business Plan

Purpose: To build a common understanding of the strategic plan and resulting annual objectives

Who Leads: CEO.

Major Activities: Strategic planning provides annual goals as an *input* to this model.

▲ Review strategies and strategic goals.
▲ Review annual objectives proceeding from strategic goals.
▲ Ensure alignment of annual objectives with strategic goals.
▲ Ensure clear metrics to measure each objective (metrics provide an *operational definition* of objectives).

Potential Tools: Discussion.

Potential Pitfalls:

▲ Nice strategic plan suitable for framing.
▲ No systematic translation of strategic and operational goals into improvement projects
▲ Resources deployed on improvement projects with little strategic leverage

Step 2: Assign Leader Accountability

Purpose: To ensure a common understanding of who is responsible for leading the achievement of each objective

Who Leads: CEO.

Major Activities:

▲ For each annual objective, determine the senior leader whose area of responsibility is most impacted.

▲ Document a time line for expected achievement.

Potential Tool:

▲ WWW matrix (who, what, when)

Potential Pitfalls:

▲ Assuming that accountability is obvious

▲ Multiple people listed as responsible with no clear leader (if everyone is responsible, no one is responsible)

Step 3: Identify Driving Metrics for Key Business Systems

Purpose: To identify the processes and systems that will have measurable impact on the identified metrics

▲ **Process definition:** Collection of work activities that transforms inputs into outputs

▲ **System definition:** Group of major processes interrelated around a specific major output

Who Leads: Executive sponsor accountable for attaining the annual objective (involve others from her or his team as appropriate)

Major Activities:

▲ Group process is led by executive (or designee).

▲ Each objective should have one or more key metrics to measure progress. The metric becomes an *operational definition* of that objective.

▲ For each metric ask, "What key business processes and systems drive this result?"

▲ Document linkage is objective—metric—process/system.

Potential Tools:

▲ Brainstorming
▲ Nominal group technique
▲ Cause-and-effect diagram
▲ Affinity analysis and interrelationship diagraph

Potential Pitfalls:

▲ Naming problems rather than processes and systems
▲ Not using good group methods (consider a trained facilitator)

Step 4: Prioritize Systems and Processes for Improvement

Purpose: To identify those processes with the highest leverage on objective achievement as measured by the key metrics.

Who Leads: Executive accountable for meeting the objective.

Major Activities:

▲ Assess each potential process against the criteria in the prioritization weighting tool (see Figure 18.5 for a sample).
▲ Prioritize based on this assessment. (Tip: The tool is there to organize the discussion and bring relevant issues to the table. The numbers are not necessarily definitive.)
▲ Finalize vital few to carry forward into improvement projects.

Potential Tools:

▲ Prioritization weighting tool
▲ Multivoting
▲ Rank ordering

Potential Pitfalls:

▲ Underestimating the difficulty of implementation, or cost
▲ Ignoring organizational "readiness" issues

Step 5: Develop Charters and Assign Resources

Purpose: To build a common, communicable understanding of the project and who is involved.

Who Leads: Executive sponsor responsible for goal attainment (or designee).

Initiative Under Consideration (Column 1)	Desired Outcome of Initiative (Column 2)	How Outcome of Initiative Will Be Measured (Column 3)	Prioritization Factors (Column 4) (Enter score in each case—see instructions below)					Priority Score/Depts (Column 5) = product of 5 previous columns & main dept affected (to ensure not overloading a dept)
			Impact on Metric (Annual Objective)	Ease of Implementa-tion	Expense	Capital Cost	Organizational Readiness	
1. Improve throughput	Better throughput	Between case turnaround time	3	3	5	5	3	675
2. Physician recruitment	More physicians doing CV	# of new phys recruited	5	2	3	5	5	750
3. Marketing process	Top of mind awareness	New patients who answer survey (because I saw your TV add)	2	5	2	5	5	500
4. Improve outcomes	Higher clinical quality	Door to balloon time	2	3	5	5	3	450

Allocate scores on a scale of 1 to 5 in whole numbers, with 1 being the lowest rating and 5 being the highest, based on desirability.

Example: A high capital cost requirement, would receive a lower rating, because high capital cost is undesirable. A high impact on metric, however, would receive a high rating.

Figure 18.5 Sample prioritization weighting tool.

Major Activities: Complete charter template.

Potential Tools:
▲ Charter template
▲ SIPOC diagram

Potential Pitfalls:
▲ Insufficient attention to constraints and nonnegotiables
▲ Poorly defined, unclear objective
▲ Poorly defined, unclear boundaries (start/stop points)
▲ Not involving people who know the process (work in the process daily)
▲ Leaving out key stakeholders, suppliers, or internal customers (adequate cross-functional involvement)
▲ Selecting inappropriate sponsor and/or leader
▲ Not involving an appropriate skilled process improvement resource (i.e., good "match" between team leader and skilled technical resource—personality, skill set, content knowledge, etc.)

Step 6: Develop Master Strategic Improvement Plan and Assess Progress

Purpose: To provide a core document that builds common understanding of the overall strategic improvement plan components, interrelated time lines, resource requirements, and accountability structure. This facilitates plan assessment, monitoring, and communication.

Who Leads: Senior leadership team.

Major Activities:
▲ Responsible executives present chosen projects and projects not chosen. Present charters for chosen projects. Build team understanding.
▲ Explore and understand interrelationships between different charters .
▲ Understand resource needs and constraints.
▲ Finalize time line and sequencing needed for each project.

TIP

Consider an off-site retreat.

▲ Summarize complete plan and assess for feasibility.

▲ If necessary, reprioritize projects, eliminate projects, and/or revise annual objectives.

Potential Tools:

▲ Use a Gantt chart.

▲ Summarize in Microsoft Project or similar tool.

Potential Pitfalls:

▲ Lack of "realism." Too many senior leadership foci—fragmented effectiveness

▲ Not seeing the interrelationships in resources and timing

Steps 6a to 6c: Monitor the Master Strategic Improvement Plan

Purpose: To develop a process for deliberate and ongoing monitoring of progress, and for adjustment as needed.

Who Leads: Senior leadership team.

Major Activities:

▲ **Develop a system for monitoring (6a):** This can be regular periodic senior leadership team (SLT) meetings, regular reports from senior executives to the COO or CEO, a multitier steering committee structure reporting to the SLT, or some combination of the above.

▲ **Carry out the monitoring system (6b):** Check progress against milestones, metrics, etc. Identify barriers to progress. Develop ways to eliminate, circumvent, or move through the barriers.

▲ **Constantly assess whether the plan as conceived is feasible (6c):** If necessary, revise the improvement plan or the annual objectives.

▲ **Continue process through achievement of the plan's objectives.**

Potential Tools: Presentation tools, brainstorming, nominal group technique, multivoting and rank ordering, structured discussion.

Potential Pitfalls: Assuming that a systematic monitoring process is not needed, lack of follow-through, becoming distracted by the "fires of the day."

Step 7: Conduct Improvement Project for Each Targeted Process

Purpose: To achieve the annual objectives through strategically targeted improvements in accordance with the charter.

Who Leads: Appropriate team leader, assisted by a performance improvement facilitator as necessary.

Major Activities:

- ▲ Choose the appropriate model and approach for improvement:
  - ▼ Just Do Its (JDIs)
  - ▼ Rapid improvement event
  - ▼ Lean Sigma project
- ▲ Identify the key customers of the process.
- ▲ Determine key customer needs. Conduct research as necessary (e.g., interviews, observation, focus groups, surveys, etc.).
- ▲ Determine baseline and targeted performance.
- ▲ Conduct improvements, redesign, or innovations as appropriate.

Potential Tools: Healing pathway analysis, value stream analysis, process and resource "leveling," visual management, 5S, control/influence assessment, priority/payoff matrix, etc.

Potential Pitfalls: Proceeding without understanding of improvement methods; ignoring constraints, inadequate resource allocation, lack of strong executive sponsorship; not involving key stakeholders, current mental models, etc.

Steps 7a to 7c: Monitor Progress on Individual Projects

Purpose: To ensure timely progress on improvements.

Who Leads: Sponsor and team leader.

Major Activities:

- ▲ **Develop a system for monitoring (7a):** This can be regular periodic meetings between the executive sponsor and team leader (with PI facilitator as necessary), regular reports from the team leader to the executive sponsor, a steering committee structure for larger cross-departmental processes made up of team leaders from individual projects and the executive sponsor (with PI facilitator as necessary), or some combination of the above.

▲ **Carry out the monitoring system (7b):** Check progress against milestones, metrics, etc. Identify barriers to progress. Develop ways to eliminate, circumvent, or move through the barriers.

▲ **Constantly assess whether the project targets as conceived are feasible (7c):** If necessary, revise the plan or the targets.

▲ **Continue the process through achievement of the project's targets.**

Potential Tools: Standard management tools, group decision-making tools.

Potential Pitfalls: Lack of adequate monitoring, project dies out due to fires of the day, lack of accountability, lack of involved executive sponsorship.

Step 8: It's a New Year—Start Again!

Purpose: To continue to align and accomplish strategic and operational objectives.

Who Leads: Senior leadership team.

Major Activities:

▲ Close the loop on this year's progress and the implications for next year's priorities.

▲ Develop new annual objectives and metrics out of the current strategic plan (or updated strategic plan, as appropriate).

▲ Translate new annual objectives and metrics into new master strategic improvement plan and carry out the plan.

Potential Tools: This model.

Potential Pitfalls: Not systematically translating strategic and operational goals into structured projects.

Future Perfect!

The first 17 chapters of this book outlined a philosophy and an approach for transforming your organization. This last chapter provides an approach to ensure that your efforts continue to mature and bear even greater fruit.

If you have read and executed the approach outlined in this book, you have no doubt come a long way toward achieving perfect care.

Congratulations to you, your leadership team, and your team members and physicians. You are providing better care to your patients, and you are doing it more effectively and efficiently.

Remember, it's about the people; it's about the process. Empower your people to continually improve your processes. May you have continued success as you strive to achieve the lofty and noble goal of perfect care.

Glossary

5S (Sort, Set in order, Shine, Standardize, Sustain)   A 5S is an improvement activity focused on cleaning up and maintaining the order of a workspace. The intent is to decrease waste, increase flow, and improve quality and productivity by minimizing errors.

5 Whys   An exercise aimed at determining the root cause of an issue by successive determination of the underlying causes of higher-order effects.

Affinity Diagram   A business tool used to organize ideas and data.

Circle Diagram   A diagram similar to a spaghetti diagram used to identify unnecessary processing. It is ideal for mapping flow of information and communication.

Continuous Flow   The sequencing of activities through the service process one task "unit of work" at a time to minimize delays and reduce the overall lead time.

Corrective and Preventive Action Program   An active organizational program for following up on corrective and preventive actions identified through the internal process audit and other methods of identifying noncompliance to operational policies and procedures.

Demand   The desire of patients, consumers, etc. for a particular commodity or service.

Deployment   The depth and breadth of the enterprise transformation across the organization. It is a measure of the maturity of the initiative.

Enterprise Map   A high-level diagram of the internal service delivery activities of an organization. Typically, each of these activities comprises a value stream/healing pathway.

FIFO (First In, First Out) Queue Queue that ensures that the order out of the queue matches the order into the queue, typically following a first come, first served (FCFS) strategy.

Flow An ideal process state in which patients, products, or materials move smoothly through the process with no delays or bottlenecks.

Functional Requirement(s) A required capability or output often accompanied by a brief summary and a rationale. This information is used to help understand why the requirement is needed and to track the requirement through the development of the system.

Healing Pathway A value stream through which patients flow. An example is the treatment of a patient in the emergency department.

Healing Pathway Analysis (HPA) A structured and facilitated event during which a cross-functional team progresses methodically through a series of interactive and evaluative steps aimed at eliminating waste, eliminating impediments to healing and improving the flow of patients through the healing pathway.

Internal Process Audit Program A systematic, documented, and ongoing review of organizational compliance to policies and procedures, including meeting the training and certification needs of all members.

Inventory Control Map A map of inventory locations indicating movement required to obtain supplies.

Just Do It (JDI) An action for which no additional study or decision making is necessary. All information and knowledge necessary to decide on the process change is available, and the team readily and easily agrees. The team has decided it should be done, and all that is left is for someone to make it happen.

Kanban System A method of using cards or other signaling devices as visual signals for triggering or controlling the flow of materials and supplies.

Lean A continual improvement methodology that focuses on improving value by eliminating waste and increasing throughput in customer-driven value streams.

Lean Sigma   A process improvement approach that combines the Lean focus on the elimination of waste and improvement in throughput with the Six Sigma focus on reducing variability and eliminating defects.

Non-Value-Added Activity   Activity performed during the production or delivery of a product or service that utilizes time or resources, but does not increase its value to the customer. There are two types of non-value-added activities. The first type comprises those activities that are non-value -added but required, due to regulatory, legal, or other requirements. The second type comprises those activities that are non-value-added and not required. This second type is also called *waste.*

Process   A set of steps that transform one or more inputs into one or more outputs.

Process Delay   The time that batches or lots must wait until the next process begins.

Productivity   The ratio of output to input. It provides information about the efficiency of the core processes.

Project   An improvement activity that entails analysis, is typically data analytic, and is often conducted by an engineer or team to determine the best approach for achieving a desired outcome.

Pull System   A delivery system in which goods are produced or services provided only when requested by a downstream process. A customer's order "pulls" a product or service from the delivery system. Nothing is produced until it is needed or wanted downstream. Compare to *Push system.*

Push System   A delivery system in which goods are produced or services provided and handed off to a downstream process, where they are stored until needed. This type of system creates excess inventory. Compare to *Pull system.*

Quality Function Deployment   A structured process that provides a means to identify and carry customer requirements through each stage of product and service development and implementation. Quality responsibilities are effectively deployed to any needed activity within an organization to ensure that appropriate quality is achieved.

Quality Management System A comprehensive, integrated, and interdependent set of policies, procedures, and checklists that is enabled by employee training, development, and certification; that is maintained by a culture of compliance including an internal process audit program, corrective action system, and personal and organizational accountability; and that leads to the reliable and repeatable provision of products and services.

Queue A line or sequence of people, equipment, tasks, and so on awaiting their turn to be attended to or to proceed.

Queuing Strategies Queuing (sequencing) techniques used to analyze and subsequently match service resources in terms of capability, capacity, and timing to demand requirements of the customer.

Quick Changeover A method of analyzing an organization's processes and then reducing the materials, skilled resources, and time needed for equipment setup. It allows an organization to implement small-batch production or one-piece flow in a cost-effective manner. Also called *Setup reduction.*

Rapid Improvement Event (RIE) aka Kaizen A facilitated, structured, team-based event aimed at further refining and actually implementing improvements to portions of a value stream.

Rapid Improvement Newspaper A tracking tool used to ensure completion of actions identified during an RIE that cannot be completed during the actual event.

Rapid Improvement Plan A plan for transforming the current state of a value stream or healing pathway to the desired future state. A rapid improvement plan will typically identify multiple Just Do Its, rapid improvement events, and projects. It will also identify both who is responsible for leading each effort and a date by which it will be completed.

Return on Investment (ROI) Profit from an investment as a percentage of the amount invested.

Root Cause Analysis A process for identifying problems, finding their causes, and creating the best solutions to keep them from happening again.

Service Generically defined as work done by one person or group that benefits another.

Service Process A service performed as part of an internal business process.

SIPOC (Supplier-Input-Process-Output-Customer) Diagram A tool used to identify all relevant elements of a process improvement effort.

Six Sigma A statistical concept that refers to the amount of variation in a process. In addition to being a statistical measure of variation, the term refers to a continual improvement methodology aimed at reducing defects.

Spaghetti Diagram A very useful graphic that shows how people or things flow in a physical layout.

Standard Operating Procedures (SOPs) Reliable instructions that describe the correct and most effective way to get a work process done.

Standard Operations The most efficient work combination that an organization can put together.

Statistical Control Measure of stability in a system. Lack of control can be caused by inconsistent adherence to policies and procedures, creating process variation that reverberates through the system, creating delays, dissatisfaction, and inefficiency.

Statistical Process Control (SPC) The use of mathematics and statistical analysis to solve an organization's problems and build quality into its products and services.

Supermarket System A stocking system in which materials are stored by the operation that produces them until they are retrieved by the operation that needs them. When a store is full, production stops.

Supply To make products or services available to satisfy a requirement or demand.

Takt Time The total available work time per day (or shift) divided by customer demand requirements per day (or shift). Takt time sets the pace of production to match the rate of customer demand. For example, if customers demand 480 x-rays per day and the radiology department line operates 960 minutes per day, the takt time is 2 minutes.

Theory of Constraints   A concept, introduced by Eliyahu Goldratt in the novel *The Goal*, that focuses on the removal of impediments (constraints) to process flow.

Total Costs   The sum of purchase price, acquisition, possession, and disposal costs. Also known as total cost of ownership (TCO) and life-cycle costs of a product or service.

Total Productive Maintenance (TPM)   A series of methods that ensures every piece of equipment in a production process is always able to perform its required tasks so that production is never interrupted.

Transformation Summit   A planning event during which the CEO and senior leadership team plan the enterprise transformation.

Transformation Team   A team under the leadership of an executive sponsor, charged to transform a value stream or healing pathway.

Value-Added Activity   Activity performed during the production or delivery of a product or service that increases its value to the customer.

Value Stream   A set of one or more processes that results in the delivery of a product or service.

Value Stream Analysis (VSA)   A structured and facilitated event during which a cross-functional team progresses methodically through a series of interactive and evaluative steps aimed at increasing value by eliminating waste and improving the flow of products or services through the value stream.

Value Stream Map   An illustration that uses simple graphics or icons to show the sequence and movement of information, materials, and actions in a value stream.

Waste   Any type of activity performed during the production or delivery of a service or product that utilizes time or resources, but does not increase value to the customer and that is not required for legal, regulatory, or other reasons.

INDEX